Technological Advances in Raising of Quality Planting Material of Tree Species in Temperate Regions

NIPA® GENX ELECTRONIC RESOURCES & SOLUTIONS P. LTD.
New Delhi-110 034

About the Editors

Prof. Arshad Hussain Mughal is presently posted as Associate Director Research at Sher e Kashmir University of Agricultural Sciences and Technology of Kashmir in Jammu and Kashmir. He completed his M.Phil in Resource Management from Indian Institute of Forest Management, Bhopal and Doctorate in Forest Seed Technology in the year 2002 from Forest Research Institute Deemed University, ICFRE, Dehradun.

He has about 35 years of experience in teaching, research and extension in Agriculture and Forestry. He has successfully guided about 15 M.Sc and Ph. D students and has published about 85 research papers in both National and International Journals. He has served as PI, All India Coordinated Research Project on Agroforestry and also PI of Network project on Wild Apricot. Besides, handled about 11 Research Projects as PI and Co-PI funded by different National funding agencies.

He has served as Programme Coordinator, Krishi Vigyan Kendra, Kargil and Scientist I/C Mountain Agricultural Research Station, Kargil, Head of the Division, Basic Sciences and Humanities Faculty of Agriculture, Wadura, Programme Coordinator, Krishi Vgyan Kendra and Extension Training Centre, Pulwama and is presently serving as Associate Director Research at SKUAST-Kashmir.

Dr. Nazir A. Pala, Assistant Professor in the Division of Silviculture & Agroforestry, Faculty of Forestry, SKUAST-Kashmir has more than ten years of experience in teaching and research. Prior to this Dr. Pala has the experience of working at UBKV, West Bengal, Himgiri Zee University Dehradun, Dolphin Institute, Dehradun, TERI, New Delhi, EIA consultant at NTPC, Govt. of India. Dr. Pala has been involved in handling several external funded projects and has more than 200 research publication and 04 books of national and international repute to his credit. Dr. Pala is the member of several international environmental organizations apart from being on the editorial board/reviewer of number of international journals.

Prof. G.M. Bhat has more than 20 years of experience in teaching, research & extension and presently working as Professor & Head Division of Silviculture & Agroforestry, Faculty of Forestry, Benhama SKUAST-Kashmir. Dr. Bhat is also OIC & PI of AICRP-Agroforestry. Prof. Bhat has been very instrumental in developing various site specifc/region specific agroforestry models for the promotion of livelihood in Kashmir Himalayas. He has also developed several revenue generating models on agro-forestry from degraded landscapes. Dr. Bhat has published more than 100 research publication including papers, books, manuals and booklets. Dr Bhat has the recognition of receiving first best teacher award in the discipline of forestry apart from receiving the gold medal in Ph. D forestry. Dr. Bhat also known by the name of green man of SKUAST-K is the national award winner for developing agroforestry model for rehabilitation of degraded lands. He is regularly engaged in conducting in plantation drives and other awareness programmes.

Technological Advances in Raising of Quality Planting Material of Tree Species in Temperate Regions

Arshad H. Mughal
Associate Director Research
SKUAST-Kashmir
Jammu and Kashmir
Kashmir, India

Nazir A. Pala
Assistant Professor
Division of Silviculture & Agroforestry
Faculty of Forestry
SKUAST-Kashmir
Jammu and Kashmir
Kashmir, India

G.M. Bhat
Professor & Head
Division of Silviculture & Agroforestry
Faculty of Forestry
Benhama SKUAST-Kashmir
Jammu and Kashmir
Kashmir, India

NIPA® GENX ELECTRONIC RESOURCES & SOLUTIONS P. LTD.
New Delhi-110 034

NIPA® GENX ELECTRONIC
RESOURCES & SOLUTIONS P. LTD.

101,103, Vikas Surya Plaza, CU Block
L.S.C.Market, Pitam Pura, New Delhi-110 034
Ph : +91 11 27341616, 27341717, 27341718
E-mail: newindiapublishingagency@gmail.com
www: www.nipabooks.com

For customer assistance, please contact
Phone: + 91-11-27 34 17 17
Fax: + 91-11- 27 34 16 16

Print ISBN: 978-93-58877-66-3
ebook ISBN: 978-93-58878-00-4

Composed and Designed by NIPA®.

Directorate of Research
Sher e Kashmir University of Agricultural Sciences & Technology, Kashmir, Shalimar, J&K 190 025

Preface

Trees are central to our lives and they are a powerful tool in mitigating green house effects. Plantations outside forest areas are the only means of reconciling the increasing demands for forest products and services on the one hand with a decreasing area of land available for forestry on the other. Good seed is one of the important components for raising a successful plantation crop. "Good seed" implies seed which is both of high viability and vigour and is genetically well suited to the site and to the purpose for which it is planted. Success of plantation programme depends mostly on high quality seeds and planting materials ensuring genetic identity and purity which significantly contribute towards sustainability. Seed is the basic input of plant kingdom; even green revolution was initiated through the seed. Seed also assumes significance in preserving genetic diversity of different tree species. The need for production of quality seed is now a day's being stressed almost everywhere as without quality seed, it is not possible to produce quality planting material. Quality planting material in the long run will help in the production of quality timber, fuel wood, fodder and other non timber forest produce.

The plant quality determines success in forest plantation as well as agroforestry establishment; particularly in relation to root development. Once tree species for plantations are selected then next important stage is to grow seedlings in the nursery that are best suited to the site conditions and also ensure that the seed stock is healthy and of good quality (i.e. it comes without pests or diseases). This can be assured by following the best possible nursery technologies available for different tree species. Besides other important factors viz. proper seed treatment, proper soil mixture, container size and type, irrigation, disease and pest management help in raising of quality plant material but are not properly addressed at the nursery stage. Technological advances in the form of raising seedling in mist chambers, root trainers and other improved containers along with use of different seed pretreatments and growing media is the best bet for raising quality seedlings in the nursery. Lack of technological knowhow not only affects the survival of seedlings in the nursery, but at a later stage plantations raised are poor and not of much economic value.

The edited book on "Technological Advances in Raising of Quality Planting Material of Tree Species in Temperate Regions" will help to address the important issues in quality seed collection and also raising of quality seedlings of temperate region in the nursery using recent time tested technologies. I compliment the contributors and hope that this manual will be beneficial to researchers, students as well as nursery managers engaged in raising of forest nurseries. I am thankful to the NIPA, New Delhi for helping in bringing out this book in the present form.

May, 2024

Editors

Contents

1

Selection of Mother Trees and Role of Maturity Indices in Quality Seed Collection

A.H. Mughal[1] and Nazir A. Pala[2]

[1]Associate Director (Research), SKUAST-K, Jammu and Kashmir, India

[2]Division of Silviculture & Agroforestry, Faculty of Forestry, SKUAST-K Jammu and Kashmir, India

---

Seed is defined as a 'matured ovule' or a reproductive unit formed from a fertilised ovule consisting of an embryo, food store and protective coat. These essential constituents are common to all seeds in different groups of seed bearing plants. The knowledge of seed morphology is essential for artificial regeneration as it can influence the collection, processing, storage and treatment of seeds. The external and internal morphologies of seeds are stable. Therefore, they provide reliable criteria for the positive identification of unknown seeds.

Importance of Tree Seed

Good quality seed is required for:

a) Planting of trees for various purposes (reduce shortage of wood and environmental degradation)
b) Execution of plantation programmes
c) Producing maximum yield
d) Introduction of exotics
e) As a source of food (nuts, fruits, berries etc.)
f) Maintaining progeny
g) Cryopreservation
h) Maintaining genetic diversity
i) Guarding against catastrophic events
j) As compact propagation material

Insufficient quantity and non-availability of quality seed supply of suitable species is often seen as a major obstacle in the development of planting programmes.

There are a number of reasons for the loss of quality genotypes. For example, people exploited bigger and high quality specimens, leaving behind smaller and poor quality ones for future generations, resulting in the gradual loss and degeneration of the genetic makeup (Lindquest, 1948). This was indicated in Japan as early as 1885 and 1896 (Tanka and Sinpo). There is every reason to ensure the production of good quality seed in sufficient quantity in order to guarantee a constant supply of genetically desirable planting stock.

Tree seed, a key input that determines the success of any tree planting activity, is often in short supply. As a result, farmers and nursery growers use whatever seed is available, regardless of its quality. In most countries, good quality tree seed is not readily available for a number of reasons. Some of them are given below:

- There is a lack of awareness concerning the importance of seed quality.
- Limited quantities of good quality seed are available.
- There are only a few areas of forests and plantations in existence that produce good quality seed.
- Seed orchards or seed stands are available for only a few species.
- The genetic quality of forests is often degraded because the best quality trees have been harvested, leaving only poorer quality trees available for seed collection.
- Collectors, dealers and other workers in the tree seed sector have limited knowledge, training material and inadequate facilities to produce, handle and store seed properly.
- No labelling or certification systems exist to provide adequate information (to the farmer and nursery grower) concerning the origin and quality of the tree seed that is available.
- No premium is paid for better quality tree seed.

Why Seed Technology?

Seed : Sexual reproduction

Seeds are unique in natural regeneration and propagation because:

1. Seeds comprise unique genetic composition, resulting from parental genetic material. The genetic variation in the offspring enhances ecological adaptability.
2. Seeds are produced in large numbers and are readily available each year or at intervals.
3. Seeds are stored for long periods under cold and dry conditions.

The science of seed biology encompasses the development and physiology of seeds until they finally germinate or fail to do so. Considering that in the context of regeneration, only one successful seed is necessary to replace the parent tree, the production of seed during its life time is excessive. For example, a fully grown eucalyptus tree may produce a million seeds or more per year and may live up to a century. In forest seed handling, we want as much of the collected seed to survive and germinate as possible. The objective of seed handling is to achieve a high survival rate of the seed.

Good Quality Seed is Required for

a) Planting of trees for various purposes (reduce shortage of wood and environmental degradation)

b) Execution of plantation programmes

c) Producing maximum yield

Seed Source

For immediate plantation programmes good quality seed and planting material can be collected from:

1. Seed production areas
2. Provenance seed stands
3. Seedling seed orchards
4. Seed trees

In case we do not have well-established seed production areas, provenance seed stands and seedling orchards for different important forest tree species, we need to opt for the identification and selection of candidate trees or plus trees/ mother trees for the purpose of seed collection, so that the resultant nursery stock is of good quality.

Candidate Tree

It is a tree that has been selected for grading because of its desirable phenotypic qualities, but that has not yet been graded or tested. It is marked with a yellow band of 5cm width, 5cm above breast height.

Superior or Plus Tree

It is a tree that has been recommended for production or breeding orchard use, after being graded. It has a superior phenotype and appears to be adaptable. It has not been tested for its genetic worth, although the chances of its having a good genotype are high. It is marked by another yellow band (2) of 5 cm width 5cm below breast height.

Elite Tree

It is a selected tree that has proven to be genetically superior by means of progeny testing. It is most suited for use in mass production of seeds or vegetative propagules.

Comparison or Check Tree

These are trees that are located in the same stand , are of nearly the same age and grow on the same or better site as the select tree and against which, the select tree is graded.

Identification and Selection of Plus Trees and Mother Trees

Most tree improvement programmes start with the selection of superior individuals within a species. Variation amongst individuals is important for a successful selection programme. If the genetic potential is poor, the performance will remain poor, regardless of environmental and silvicultural efforts. If the genetic potential is good, it may be expressed by appropriate silvicultural measures. In the selection of seed sources and seed trees of unknown genetic constitution, a few measures and precautions can and should be taken, in order to avoid genetic inferiority. These include avoiding seeds from related individuals or inbred populations and also, phenotypically inferior trees. The characteristics of trees for various end uses will differ. Even for each end use, they may differ depending upon whether the trees are grown in a plantation or under agroforestry conditions. Thus, improvement is based on the objective to be achieved.

Criteria for Selection of Timber Trees

Straight bole, fast growth, less forking, thin branches at right angles, small crown, thin bark, high specific gravity, long fibre length, less flowering, tolerance to disease and pests.

Criteria for Selection of Fuel Wood Trees

Fast growth, excessive branching, medium to high density, thornless, adaptability to diverse ecological conditions, high coppicing power etc.

Criteria for selection of fodder trees

Short height, more branching, ability to withstand high lopping, quality fodder, tolerance to adverse conditions.

Methods of Plus Tree Selection

1. **Comparison tree method:** The comparison tree check method is generally employed in even aged forests where the crop is uniform.

During the selection, observations are recorded on the most important economic characteristics. 5-10 trees of similar age are selected from the dominant crown. The trees selected should be about 50 to 100 m apart. The candidate tree is selected from amongst these selected trees. A candidate tree is selected as a plus tree if it proves its superiority over the comparison tree; otherwise, it is rejected.

Pro Forma for Candidate Tree Report

Species: Candidate Tree No.

Location: Site quality:

Date: Site Address:

Report by: Organisation:

Characteristics of economic importance for timber production

Characteristics	Comparison tree				Average	Candidate tree	Superiority (%) of candidate tree over average
	1	2	3	4			
General Health (G, Vg, E)							
Age of tree							
Height (m)							
Diameter (cm)							
Taper							
Forking spread (m), Height							
Crown Width (m) Length (m)							
Branches Angle Diameter (L,M,U)							
Seed bearing (H, M, L)							
Disease/ Insect / Pest attack (A, M, N)							

Scoring measurements (G, Vg, E= Good, Very Good and Excellent); (L, M, U = Lower, Medium and Upper branches); (H, M, L= High, Medium and Low); (A, M, N= Abundant, Medium and Nil)

Source: (Khajuria et al.,2003)

2. **Regression method:** This method is used in uneven or mixed forest stands. It requires the development of regression lines for a stand or site, relating the expression of characters to the tree age by sampling a number of trees. The sampling requires trees of different ages. The candidate trees are compared to regression lines. The candidate trees which meet the minimum selection standards above the regression lines are selected as plus trees, while the others which fall below the regression line are rejected. It is important to note that the regression lines are site and species-specific. This method cannot be used in species where the age cannot be ascertained. (Repetition).

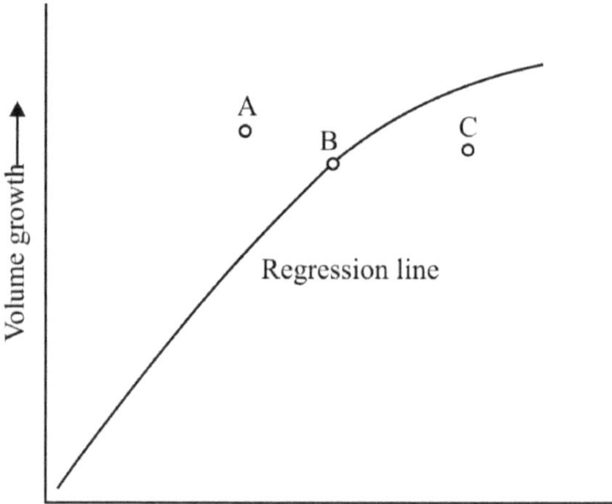

Role of Maturity Indices in Quality Seeds Collection

The knowledge of the exact stage and time of seed maturity is essential for the collection of an abundant quantity of healthy and vigorous seeds, especially in conifers, where good seed years occur after many years, viz., once in 4-5 years in Deodar (Troup, 1921) and once in 2-6 years in *Picea* (Pearson, 1931). Besides, seeds that are collected when fully ripe retain viability longer than those collected when immature (Harrington, 1970, Stein *et al.*, 1974). Immature seeds are low in viability and often produce low vigour, deformed seedlings (Heit, 1961; Schubest, 1956). Thus, fruit collection should be started only when the seeds are sufficiently mature. Therefore, the indicators of maturity for individual species are significant so that collection is made at the right time; otherwise, the collection of immature seeds will result in loss of time, money and failure of plantations. Similarly, collection after dispersal is difficult as it is time consuming and not much seed from identified mother trees can be collected. Therefore, it is prudent on the part of the collector to collect seeds when they are sufficiently mature.

Colour Change

Colour changes in the fruit or cone provide a simple and in some species, reliable criteria for judging seed maturity (Willan, 1985).They are also the easiest to use in the field. Colour changes usually range from green of the immature fruit to various shades of yellow, brown or grey and this may be accompanied by the hardening of the pericarp of dehiscent woody fruits. The changes in fruit colour is, by far, the best available criteria for judging the maturation of seed, as shown in *Carissa opaca* and *Ficus benjamina* (Maithani *et al.*, 1987ab), where it was directly linked with a change in fruit colour. Among conifers, the change in fruit colour can also be a useful index of ripeness. For instance, in juniper (*Juniperus*) fruits, the change to deep blue colour is a good index of ripeness (Stoeckeler and Slabaugh,1965). A purplish colour in balsam fir (*Abies balsamea (L)*:Mill) cones suggest maturity (Stoeckeler and Jones,1957). Among pines (*Pinus strobus.L*), seeds are mature when the cones turn yellowish-green with brown on the scale tips; red pine (*P. resinosa Ait.*) cones are ripe when they turn purplish with brown on the scale tips; jack pine (*P. banksiana Lamb.*) cones are ripe when half or more of the cone surface is brown (Stoeckeler and Jones,1957) and *P. clausa* cones are mature when they turn brown in colour (Barnett and McClemore,1965). Cone colour of *Cedrus deodara* change from green in the beginning to dark brown with some green patches at maturity (Mughal and Thapliyal, 2006). *Quercus robur* acorns change from green to chocolate brown at maturity (Bhat *et al.*,2007). Bhat and associates (2017) reported that cone colour in *Pinus wallichiana* changed from light green to green and finally, green with brown patches at maturity in the Kashmir Himalayas. Similarly, seed colour changed from whitish to light brown and finally, dark brown at maturity. Bhat (2017) reported that cone colour of *Abies pindrow* changed from pinkish to light chocolate and finally, deep chocolate brown at maturity in the Tangmarg and Sind Divisions in Kashmir region.

Specific Gravity

Specific gravity or the float test method is another reliable and quick method for determining the maturity of fruit or cones in the field. Specific gravity or density is the ratio of the unit weight of the fruit or seed to the unit weight of the volume of water displaced by the fruits or seeds. It decreases in the same manner as the moisture content decreases with the maturation of fruit and cones. Float tests should be performed quickly and five or more cones should float before the crop from an individual tree is labelled as 'mature' (Barnett, 1979). Specific gravity indices of maturity have been established for a number of species by various authors. A specific gravity of 0.85 at maturity for *Abies concolor*, 0.90 for *A. grandis* and 0.70 for *A. magnifica* have been worked out by Oliver (1970) and Rediski and Nicholson (1964). Mughal and Thapliyal (2006) reported that the specific gravity of *Cedrus deodara* decreased from 1.0 to 0.78 at maturity in North Kashmir and 1.0 to 0.66 at maturity in the Jammu region. Germination percentage was reported to increase with the decrease in specific gravity. This relationship between declining cone specific gravity and increasing seed germination has also been reported by a number of workers in different tree species (Maki, 1940; Fowells, 1949; Eliason and Heit 1954; Wakeley, 1954; Pfister, 1967; Oliver, 1974 ; Rietveld, 1978). In *Abies pindrow*, a specific gravity of 0.97 at maturity has been reported by Singh (1998). It was observed that germination also increased as the cones proceed towards maturity. The specific gravity of cones decreased from 1.04 in August to 0.97 in October. The specific gravity in *Pinus wallichiana* cones decreased from 1.13 to 0.90 as the cones proceeded towards maturity, as reported by Bhat and associates (2017) in the Kashmir Himalayas. Specific gravity recorded a decrease in *Abies pindrow* and it was between 0.76 to 0.70 at maturity at different altitudes in the Tangmarg region of Kashmir. In Sind Division, it was between 0.79 to 0.74 at maturity at different altitudes (Bhat, 2017).

Seed Contents

Seed contents can also be examined for maturity by cutting fruits or cones lengthwise. Mature seeds have a firm white endosperm and a fully developed firm embryo. Immature embryos and endosperm pass through an immature 'milky' stage, followed by a 'dough' stage, when the tissue becomes firm (Prasad and Kandya, 1992). Since the endosperm is already present when fertilisation occurs, the post fertilisation maturation of most gymnosperm seeds centre on embryo development. The embryo grows and differentiates within the endosperm into a miniature plant with a rudimentary root or radicle, stem or hypocotyl and bud or plumule with cotyledons. The seeds of the conifers of Western North America are mature enough for collection when a cutting test shows that their contents have changed from a milky and viscous condition to a firm and crisp consistency, similar to the meat of a coconut (Dobbs *et al.*, 1976). Embryo size has also been used as a maturity index. The length of the embryo in the seeds of the conifers of Western North America should be at least 75% of the length of the embryo cavity (Dobbs *et al.*, 1976).

Seed Germination

The percentage of seed germination also increases as the cones mature. Many authors have correlated seed germination percentage with the maturity of seeds. Mughal and Thapliyal (2006) reported that in North Kashmir, germination was 0% at the first collection in the month of August, but subsequently increased to 13.25% and then to 59.50% at maturity. In South Kashmir germination increased from 6.55% in August to 75.50% at maturity in the month of September. Similarly, in the Jammu region, there was an increase from 1% in August to 46.25% and 67.50% in the month of September and October respectively. Studies conducted on seed germination and seedling establishment in the natural forests of spruce and silver fir in Kotgarh Forest Division, Himachal Pradesh, showed that in both species, the seeds dispersed in October had a lower germination percentage than those dispersed in November (Singh and Singh, 1984). This pattern of germination can also be correlated with specific gravity i.e., with a decrease in specific gravity, there is an increase in seed germination. The relationship between decreasing cone specific gravity and increasing seed germination has also been reported in different tree species (Maki, 1940; Oliver, 1974; Rietved, 1978). Singh (1989) has also reported a similar trend in spruce cones with the germination percentage increasing from 0.70 in August to 48.6 in October at maturity. Singh (1998) obtained similar results in *Abies pindrow* cones, with the germination percentage increasing from 0.50 in August to 32.14 in October at maturity at 2,600m elevation. Similar trends were observed at 2,850m elevation. In *Pinus wallichiana*, the germination percentage increased with maturity at different altitudes. At Kehmil Forest Division in Kashmir, germination increased from 13.71% to 67.37% at maturity, while in Lidder Forest Division, it increased from 25.47% to 70.50% at maturity (Bhat *et al.*, 2017).

References

Barnet, J.P.(1979). Maturation of tree seeds. In:Proc. Flowering and seed development in trees. A symposium : 272-280. South For. Exp. Stn. Miss. State Univ. ,IUFRO

Barnett, J.P. and McLemore, B.F.(1965). Cone and seed characteristics of sand pine. USDA Forest Service Research Paper, SO-19.

Bhat, G.M.; Mughal, A.H.; Malik,A.R; Khan, P.A; Sofi,P.A and Islam, M.A (2017). Cone and seed maturity indices in Pinus wallichiana under temperate conditions of Kashmir Himalayas, India. Journal of Applied and Natural Science 9(4);1987-1993.

Bhat, G.M.; Khan, M.A. & Mughal,A.H.(2007). Maturity indices for harvesting Quercus robur acorns under temperate conditions of Kashmir valley. Eco. Env. & Cons. 13(4); 2007, 741-743.

Dobbs,R.C., Edwards,D.G.W., Konishi,J and Wallinger,D.(1976). Guidelines in collection of cones of British Colombia conifers, B.C.For. Ser. and Ca. For. Ser. Joint Report No.3.

Eliason,E.J and Hill,J.(1954). Specific gravity as a test for cone ripeness with red pine. Tree Plant. Notes No. 17, pp1-4.

Elliason, E.J. and Heit, C.E.(1973). Red pine seed shows high germination after 42 years in storage. J. For., 71:776.

Fowells, H.A.(1949). An index of ripeness for sugar pine seeds. USDA Forest Service. Forest and Range Experiment Station Research Note, 64:1-5.

Harrington, J.F.(1970). Seed and Pollen storage for conservation of plant gene resource. In: Genetic resources in plants, their exploration and conservation, Handbook, No. 11, International Biological Programme, London.

Khajuria, H.N; Sidhu,D.S & Chauhan,S.K (2003). Tree Improvement Practical Manual. Department of Forestry and Natural Resources, PAU, Ludhiana. 141 004.

Lindquest,J.F.(1948). Genetics in Swedish forestry practices. Stockholm pp 173.

Maki,T.E.(1940). Significance and applicability of seed maturity indices for Ponderosa pine. J. Forestry., 38:55-60.

Mughal, A.H. and Thapliyal, R.C (2006). Cone and seed maturity indices in of Cedrus deodara. Indian Journal of Forestry. 29(2) :167-174

Oliver, W.W.(1970).Data filed Apr.13,1970. USDA. Forest Service, Pac. Southwest Forest and Range Exp, Stn. Redding, Calif.

Oliver, W.W.(1974). Seed maturity in white fir and red fir. USDA Forest Service Research Note, PSW-99:1-12.

Pearson,G.A.(1931). Forest types in the South west as determined by climate and soil. U.S.Deptt. Agri. Tech. Bul. 247, pp144.

Pfister, R.D.(1967). Maturity indices for grand fir cones. USDA Forest Service Research Note, INT-58:1-7.

Prasad , R and Kandya,A.K .(1992). Handling of forestry seeds in India. Associated PublishingCompany. N. Delhi.

Rediske, J.H and Nicholson,D.(1964). Maturation of noble fir seed- a biochemical study, Weyerhaeuser. For. Pap., 2, pp15.

Rietveld,W.J.(1978). Forecasting seed crops and determining cone ripeness in south-westernPonderosa pine USDA. For. Ser. General. Technical Report R.M 50:1-12.

Singh,O.(1998). Seed maturity indices in Silver fir (Abies pindrow Spach.). The Indian Forester., 124(3):243-246.

Singh,V and Singh,R.V.(1984). Seed dispersal, seed germination and seedling establishment in natural forests of silver fir and spruce. II Seed germination and seedling establishment. The Indian Forester, 110:632-639.

Singh,V.(1989). Seed maturity indices in Spruce- The Indian Forester, 15(5):342-347.

Stein, W.I., Slabaugh, P.E. and Plummer, A.P.(1974). Harvest, processing and storage of fruits and seeds. In: Seed of Woody Plants in the United State, Agriculture Handbook No. 450, USDA Forest Service, Washington D.C. USA, 300-320.

Stoeckler, J.H and Jones,G.W.(1957). Forest nursery practice in the Lake States. U.S. Dept. Agric Handbook 110, pp124.

Stoeckler, J.H and Slabaugh, P.E.(1965). Conifer nursery practice in the Prairie- plains. U.S.Dept. Agric. Agric Hand Book. 278,pp93.

Tanaka, Z. (1885). Species of good timber will become extinct in natural forests. Dainippon Sanrinkai Hookoku. Bull. Jap. For. Assoc., 38:135-142. Japan.

Troup, R. S. (1921). The Silviculture of Indian Trees. Vol. III.pp1013-1095 Clarendon Press. Oxford.

Wakeley, P.C.(1954). Planting of southern pines. USDA, Agriculture Monograph,18:1-233

2

Effect of Cold Stratification on the Germination of Seeds of Conifers from Indian Himalayan Region

Shalu Devi Thakur

Research cum Facilitation Centre, North. National Medicinal Plant Board, Ministry of AYUSH, Faculty of Agriculture, Wadura, SKUAST-Kashmir, Jammu and Kashmir

Introduction

Germination is a critical stage for the establishment of plant species, especially those that grow naturally in arid and semi-arid environments (Alouani and Bani Aameur, 2004). Several factors may influence germination percentages in natural environments. Among these factors, seed dormancy contributes in part to germination failure of many species under adverse environmental conditions (Bewley, 1997). There are two categories of dormancy: coat-enhanced dormancy (seed coat dormancy), where the embryos isolated from these seeds are not dormant and embryo dormancy (internal dormancy), where the embryos themselves are dormant (Bewley, 1997; Ellery and Chapman, 2000). Seed dormancy is reported to be an adaptive mechanism that ensures the survival of the species through periods of environmental stress (Gutterman, 1993; Bell *et al.,* 1995; Baskin and Baskin, 1998; Gusano *et al.*, 2004; Kermode, 2005). Many unpredictable factors can control the intensity of this dormancy and lead to high degrees of intra-specific variation at several levels (Andersson and Milberg, 1998). The environmental conditions of the maternal plants during seed maturation are reported to be an important factor in controlling seed dormancy and germination, accounting for a significant proportion of the variation in germination percentage and are likely to be responsible for the differences in germination between seeds from different years (Andersson and Milberg, 1998). Conifer seeds, in general, have a high degree of dormancy, even if subjected to environmental conditions favourable for germination (Jull and Blazich, 2000). This dormancy is caused by a combination of internal (physiological) and external (physical) factors (Basu, 1994). Hard seed coats, together with pericarps and other structural barriers, offer a high mechanical resistance and this acts as a barrier for the imbibition of water and exchange of gases, which is essential for the initiation of the germination process (Kelly, 1992).

Seeds in Conifers

A seed has been defined as a 'mature ovule' or a reproductive unit formed from fertilised ovule. Seeds are essential in the regeneration of forests. It is the primary method used by conifers to reproduce, maintain genetic variability and become established on appropriate sites. A conifer seed has three main components: seed coat, embryo and mega-gametophyte.

Seed Coat

It protects the inner structure of the seed from damage, but can also restrict oxygen uptake, gas exchange, water uptake or radicle emergence due to its anatomical structure. It is the readily observable feature of the seed and protects the inner tissues from insects, fungi, adverse conditions and mechanical damage. True firs *(Abies spp.)*, hemlocks, and western red cedar have resin vesicles in their seed coats (Singh, 1978). These vesicles are surrounded by epithelial cells that produce and secrete resin into the vesicle. Resin vesicles form in the early stages of seed development in the middle or outer layer of the seed coat and are usually more abundant on the lower surface of the seed (formerly in contact with the ovuliferous scale).

Embryo or Zygote

After the fertilisation of an ovule, the embryo develops to maturity by a complex series of stages known as embryogeny. All the rudimentary structures and information necessary to produce a mature tree are contained in the embryo. The embryo is the only seed component containing genetic information from both parent trees.

Mega-gametophyte

The mega-gametophyte surrounds the embryo, thereby providing protection and nourishment for initial growth. It originates from the mother tree and is haploid. The mega-gametophyte cells are large, thin-walled and spherical in shape. In Douglas Fir, the dry weight composition of the mega-gametophyte is 60% lipids, 16% proteins and 2% sugars (Owens, *et al.*, 1993). Lipids are an efficient means of energy storage due to the larger number of carbon- hydrogen bonds that release a greater amount of energy when oxidised than other organic compounds (Slack and Browse, 1984).

Seed Handling System

The seed handling system encompasses all seed handling activities from cone collection to sowing in the nursery. The system begins with the collection of cones from seed orchards, wild stands or plantations. Post-collection handling, including temporary storage, monitoring and transport of cones to a processing facility, is a key step in the production of high quality seeds. It is generally recommended that the cones of most species should be field-stored (interim storage) for approximately four weeks prior to shipping to the extractory in order to reduce moisture content and risk of damage. Cones are generally opened through a kilning process and the seeds extracted by tumbling or screening. Cones of *Abies spp.* should not be kilned as they naturally disintegrate with additional conditioning. Cone processing includes the drying process and extraction of viable seeds from cones.

Seed Collection

Cone collection and the post-collection handling of cones are very important in determining seed lot quality. Several aspects are worth emphasising here. The first step is locating stands that meet your demands followed by monitoring the crop throughout cone development. One should have thorough knowledge about cone and seed maturity, potential yield and degree of pest activity before collecting cones. Sampling should become more frequent as the cones and seeds approach full maturity (generally, August to September) (Eremko *et al.*, 1989). Seed collection requires good planning in advance regarding the deployment of trained staff, arrangement of transportation facilities, seed collection equipments, measures to ensure the safety of the workers, packing and labelling material, maintenance of the records, etc. Cones from natural stands and seed orchards may be collected using any of several methods, including helicopters, climbing, felling, squirrel caches, etc. The method of cone collection and its efficiency will vary with species and crop intensity.

Seed Dormancy

Dormancy is considered a mechanism that eliminates the risk of germination in the autumn after seeds have been released from their cones. Seed dormancy is overcome through the technique of imbibitions stratification (moist chilling), which exposes the seed to cool (2–5°C) temperatures for a specific duration following the imbibitions of the seed. There are mainly two types of seed dormancy - physical and physiological. In physical seed dormancy, the seed possesses anatomical features that either restrict the entry of substances, such as water and oxygen or restrain the emergence of the radicle. The physical restraint of the seed coat accounts for the majority of the dormancy exhibited by some pines (Barnett, 1976). The thickness of the seed coat of pines inhibits germination (Barnett, 1991). Physiological dormancy, also called embryo dormancy, is not well understood, although it is widespread among conifers Embryo dormancy is generally controlled by a balance of hormones that inhibit germination and those that promote it (Khan, 1975), although changes in tissue sensitivity to these hormones may be more important (Trewavas, 1981). For the effective removal of physiological dormancy, an optimal moisture content ranging from 30% to 35% exists (Edwards, 1982).

Dormancy in Conifers

Quiescence is a common property which plays a significant role in promoting or delaying the germination of seeds until environmental conditions are favourable for it. In the presence of favourable conditions, certain seeds fail to germinate because of dormancy. Conifer seeds generally have intermediate dormancy and this is overcome by cold stratification (Nikolaeva, 1969). Seeds with intermediate dormancy respond to chilling. The seeds of many species fail to germinate despite the presence of favourable environmental conditions. Janick (1974) termed this state as seed dormancy. Apart from environmental factors, the genetic component of the species also causes dormancy. From the forester's point of view, dormancy has some disadvantages. Delayed and irregular germination in the nursery is a serious impediment to efficient nursery management (Bonner *et al.*, 1974).

Stratification

Stratification (imbibition followed by moist chilling) has mainly been described as a method to overcome embryo dormancy, but stratification may overcome coat-induced dormancy as well as offering other benefits. Many studies have indicated that stratification will improve the speed and uniformity of germination, which are important considerations in producing a uniform seedling crop. The effectiveness of stratification in overcoming seed dormancy requires that certain conditions be met: an appropriate moisture level, temperature, duration and access to oxygen.

Cold Stratification

Cold stratification is the most common method employed to break seed dormancy and ensure the uniform and quick germination of seeds in many forest tree species (Barton, 1951; Haavisto and Winston, 1974). It has been widely used as a pre-sowing treatment for breaking dormancy to enhance the seed germination rate (ISTA, 1976; Baskin and Baskin, 2004). This is an effortless, cheap and successful method for overcoming seed dormancy. The effects of moist chilling in establishing hormonal levels in seeds have been proved due to the initiation of appropriate enzyme activity (Nikolaeva, 1969). Moreover, the phenomenon of cold stratification has long been recognised in overcoming the physiological dormancy of seeds of many species (Baskin and Baskin, 1987).

Moist chilling breaks the dormancy and accelerates the rate of germination in physiologically dormant seeds (Wang, 1987). Moist chilling of dormant seeds may generally be efficacious, particularly if damage has accumulated due to natural deterioration or as a result of an imposed accelerated ageing regime (Mittal, 1987).

Young and Young (1992) recommended pre-chilling at 3-5°C for 14 days for the germination of deodar seeds. The seeds of *Cupressus torulosa* germinate very poorly in the laboratory as well as field conditions (Schompeyer, 1974; Young and Young, 1992). Cold stratification for three weeks can improve the germination rate in *Cupressus torulosa* (Thompson and Morgan, 1990). Pre-chilling, commonly referred to as cold stratification, has often been used to accelerate the germination of the seeds showing internal or physiological dormancy (Tamta *et al.*, 2001). Malik *et al.,* (2009) observed that due to the poor nursery performance of the *Pinus gerardiana* seeds, a stratification period of 60 days under 4 ± 1°C temperature significantly improved germination (64.35%). Pawak (1993) also reported better germination and seedling growth in Alaska red pine seeds that had been stratified for 60 days under nursery conditions.

Some conifer seeds may require one month or less of cold stratification to emerge from dormancy. For example, *Abies balsamea* requires 28 days (Franklin, 1974), *Abies densa,* 28 days (Beniwal and Singh, 1989) and *Pinus contorta*, 21 days (Hassis and Thrupp, 1931). Others such as *Pinus albicaulis* may require 90 to 120 days, while still others like *Pinus cembra* may require 90 to 270 days of stratification (Krugman and Jenkinson, 1974) to overcome dormancy. *Cedrus deodara* (D.Don) seeds have been reported to respond to chilling and 21 days pre-chilling is recommended by ISTA (ISTA, 1966, 1976). In *Cedrus deodara,* higher germination percentage has been reported as a result of chilling of seeds (Thapliyal and Gupta, 1980). The length of stratification required depends upon the degree of dormancy, which in turn, is influenced by seed source (Bonner *et al.*, 1974). The optimum stratification temperature falls between +2 and +5°C.

Temperatures below freezing point should be avoided as they are ineffective in breaking dormancy and may also injure the imbibed seeds. Stratification at 5°C resulted in faster germination than at 2°C for *Ponderosa pine* and Douglas fir. However, if the higher end of this range is chosen, one should frequently monitor the seeds as germination under stratification conditions is possible (Downie *et al.*, 1998). Trial work on 10 seed lots of lodgepole pine and interior spruce, stratified at moisture contents between 15% and 45%, indicated that maximum germination occurred at 30% moisture content for both species (Hannam. 1993). The optimum stratification temperature falls between +2 and +5°C. Stratification at 5°C resulted in faster germination than at 2°C for *Ponderosa pine* and Douglas fir, but if the higher end of this range is chosen, one should frequently monitor the seeds as germination under stratification conditions is possible (Danielson and Tanaka, 1978).

Sofi and Bhardwaj (2008) find that the germination percentage and germination value increased with the increase in stratification period up to 75 days and decreased thereafter. Maximum germination of 82.10% and germination value of 19.50 were recorded in the seeds stratified for 75 days and they recorded minimum values of 57.10% and 9.86 respectively in the seeds which were not stratified (control). The results pertaining to the study conducted by Rawat *et al.*, (2008) revealed the effects of GA3, moist–chilling and temperature on seed germination of *Abies pindrow* and *Picea smithiana* from five different provenances. Seeds were soaked in GA3 (10 mg/L-1) for 24 hours and then chilled at 30 -50° C for 15 days. Four temperature regimes, viz., 100° C, 150° C, 200° C and 250° C were used for stimulating seed germination. Results showed that soaking and chilling significantly increased the germination percentage. The germination percentage was highest at 100° C. Overall results showed that soaking seeds in GA3 (10 mg/L-1) for 24 hours, moist chilling for 15 days and germinating at 100° C produced effective germination in both the species studied.

For enhancing the rate and percentage of germination and for breaking the dormancy, moist chilling or cold stratification has been widely used as a pre-sowing treatment (Schopmeyer, 1974; AOSA, 1992; ISTA, 1999; Wang and Berjak, 2000). This is a simple, inexpensive and effective technique for overcoming seed dormancy. Though the phenomenon is not yet fully known, the effects of moist chilling in establishing hormonal levels have been proved due to the initiation of appropriate enzyme activity (Nikolaeva, 1977). Gibberellic acid (GA3) has been shown to promote the germination of seeds (Vogt, 1970; Krishnamurty, 1973; Chandra and Chauhan, 1976; Ghildiyal, 2003). The germination percentage in the seeds of *Nothofagus obliqua* increased when pre-chilled after soaking in GA3 solutions for 24 hours (Shafiq, 1980). Singh (1973) reported that spruce seeds germinate comparatively more profusely than those of silver fir and every year, enough seeds become available for raising sufficient planting stock. Singh *et al.* (1975) have recorded 56% germination of spruce seeds in B.O.D. incubator and maximum 39.5% germination in pot culture experiments. On the other hand, pre-chilling or cold stratification has been found to be an effective method for germinating the seeds of many other species (Shafiq and Omer, 1969; Miller, 1971; Stilinovic and Tucovic, 1971). Work on seed testing of various provenances of *Pinus roxburghii* from Uttarakhand and Himachal Himalayas has been done by Sharma *et al.* (2001), Ghildiyal *et al.* (2007, 2008, 2009) and Ghildiyal and Sharma (2005, 2007). Studies on both physiologically dormant *Picea glauca* and non-dormant *Picea mariana* (black spruce) seeds have shown that moist chilling was beneficial in accelerating the rate of germination (Wang, 1987). Moreover, the phenomenon of cold stratification has long been recognised in the process of overcoming physiological dormancy of the seeds of many species. The work of Wang (1973 and 1987) on *Pinus* and *Picea* species indicates that short periods of moist chilling of such non-dormant seeds may generally be efficacious, particularly if damage has accumulated due to natural deterioration or as a result of an imposed accelerated ageing regime. The hypothesis, therefore, is that the repair mechanism may be activated during moist chilling in non- dormant, cold temperate gymnospermous seeds, which are naturally aged or have been subjected to accelerated ageing. If this is generally applicable, cold stratification could have far-reaching implications for nursery practices, at least for such gymnosperms, in which, aged and stored seed lots are used. In the present study, we have tried to explore the effect of cold stratification on the seeds of *Pinus roxburghii*, taking into account various physiological processes in relation to post-harvest desiccation and chilling requirements.

Results of Various Studies That Have Been Taken Place in India

- Mughal and Thapliyal (2006) observed an increase in the germination percentage in the seeds of *Cedrus deodara* towards the cone maturity. It increased from 0.02% in August to 59.50% at maturity in October in North Kashmir. In South Kashmir, the germination percentage increased from 6.55% in August to 75.50% at maturity in September, while it increased from 1.0% in August to 67.50% at maturity in the month of October in Jammu. They concluded that it was the result of the loading of seeds with carbohydrates, fats and proteins, which proceeds gradually across the season to maturity, a process commonly known as germinability.
- The study carried out by Sofi and Bhardwaj (2008) concludes that seed weight has a significant effect on the germination and seedling growth parameters. They found that medium-sized seeds (0.10-0.22g/seed) resulted in the maximum germination of 74.20%, followed by large-

sized seeds (>0.22g/seed) with 70.32% and the least in small-sized seeds (<0.10 g/ seed) with 57.50%. They also reported the same trend in the seedling growth parameters where plant height was 19.24 cm, collar diameter was 2.73 mm, shoot-root ratio was 3.06 and the total biomass was 3.82 g in the case of medium-sized seeds. Sowing dates also have a significant effect on germination and other nursery parameters of the conifers. There are various studies which show that the germination of conifer seeds in the nursery are influenced by sowing dates.

- Mughal *et al.*, (2007) discovered the effect of various sowing dates on the germination parameters of *Pinus wallichiana* and *Cupressus torulosa.* They recorded a maximum germination of 92% in *Pinus wallichiana* in seeds sown during two fortnights in February (winter). However, it was at par with the germination of seeds sown in December. In case of *Cupressus torulosa,* they recorded a maximum germination of 98% in the seeds sown during December, February and the first fortnight of March. They reported a survival percentage of 86% in *Pinus wallichiana* for seeds sown on 1st February. Though germination was at par with that on 1st March, they reported a survival percentage of 98% in the case of seeds sown during the second fortnight of February for *Cupressus torulosa.* From this, they concluded that the survival percentage is affected by the fact that the temperature during December and January is below freezing point, which probably damages the seeds as well as the small germinations, as a result of the incidence of frost. They further concluded that the insect and pest activity during the spring season causes more damage to the seedlings. A period of 165 days was taken to complete germination by the seeds which were sown in autumn and 65 days by those sown in summer, in the case of *Pinus wallichiana.*
- The study titled "*Effect of different pre-sowing treatments on seed germination of spruce* (*Picea smithiana Wall. Boiss*) *seeds under temperate conditions of Kashmir Himalayas*" was carried out in a laboratory at the Faculty of Forestry, Sher-e-Kashmir University of Agriculture, Science and Technology of Kashmir, Benhama, Ganderbal, Jammu and Kashmir, during 2011-2012, to investigate the effect of different GA3 concentrations on the germinability and growth of *Picea smithiana* seeds under laboratory and nursery conditions, which were imbibed for three different durations. In the present investigation, good quality spruce seeds were soaked in water for 24 or 48 hours and also given GA3 treatment at 50, 100, 150, 200 or 250 ppm. Control seeds

were not soaked or applied GA3. A total of 13 treatment combinations were evaluated for their impact on seed germinability parameters under laboratory and field conditions and seedling growth parameters under field conditions.

Seed Germination Parameters

Laboratory Conditions

Gibberellins are the naturally occurring plant growth hormones. GA3 treatment can overcome dormancy in different seeds that have a hard seed coat or dormant embryo. In most species, the survival percentage, growth and total biomass increased when the seeds were pre-treated with GA3. The result of the influence of gibberellic acid on seed germinability is presented. Germination percentages of *Picea smithiana* seeds, with or without soaking in GA3 over varied periods, differed significantly ($p \geq 0.05$).Without GA3, the germination percentage of the seeds was low and started late. In contrast, when the seeds were imbibed with different concentrations of GA3 for varying durations, the germination percentage rose to 39.50% without soaking (control) to 75.50 % when seeds were soaked in 200 ppm. GA3 for 48 hrs. Soaking in distilled water and lower concentrations of GA3 was ineffective in breaking fully the dormancy of the seeds to produce the maximum germinability of the viable seeds, indicating that *Picea smithiana* seeds have physiological dormancy. Similar trends were observed for the other germination parameters, viz., germination capacity (85.50 %), germination energy (55.46), germination speed (32.89) and germination value (10.58) when the seeds were soaked in 200 ppm GA3 for 48 hrs. They differed significantly from the seeds which were soaked for 24 hours in 200 ppm GA3 and recorded a germination percentage of 70.62 %, germination capacity of 80.75 %, germination energy of 48.99, germination speed of 28.52 and germination value of 5.90. The minimum germination parameters were recorded in the seeds which were sown without any treatment (control). The germination parameters increased with the increase in the GA3 concentration and soaking duration up to 200 ppm and 48 hours respectively and decreased with further increase in the GA3 concentration and soaking duration.

Nursery Conditions

The germination of the seeds treated with different GA3 concentrations revealed that germination parameters, viz., germination percentage and germination energy increased and varied significantly at $p \leq 0.05$ from 30.11% and 10.58% (control) to 64.00 to 34.21% (200 ppm GA3 for 24 hours) respectively. Similarly, the germination value and (plant percent?) increased linearly and significantly at $p \leq 0.05$ from 0.75 and 19.90 % under control and reached the maximum of 2.82 and 54.50 respectively, when the seeds were treated with 200 ppm GA3 for 24 hours. They decreased gradually after a further increase in the GA3 concentration and duration of imbibition under field conditions.

- The increase in spruce seed germination by optimum concentration of GA3 might have been due to the enhancement of hydrolase (especially δ-amylase) synthesis, as reported by Paleg (1960a and b; Amen 1968) or probably due to the first initiation of embryo growth and subsequent synthesis of more GA3 that might have induced hydrolase synthesis (Chen and Varnes, 1973). Several studies have shown gibberellins to be an effective germination stimulator (Sofi, 2005; Lavania *et al.*, 2006).
- An increase in the germination of chilgoza pine seeds by increasing soaking periods was probably attributed to the enhancement of hydrolase (especially amylase) synthesis, as reported by Bewley and Black (1994) and Chen *et al.* (2008).
- Cold stratification followed by GA3 application has been found to suppress this inhibition and enhance germination (Mcbridge and Dickson, 1972) in spruce seeds. Shivani (2003) observed an increase in the germinability of *Abies pindrow* seeds after 24 hours of soaking in water at 2°-3°C under laboratory conditions and 48 hours of water soaking at 2°-3°C under field conditions, followed by 200 ppm GA3, soaking for 24 hours.
- In *Picea smithiana,* soaking of seeds for 24 hours at 2°-3°C + 100 ppm GA_3 application increased germination under laboratory conditions, but under field conditions, 48 hours soaking in water was better.
- Lavania *et al.* (2006) observed that for higher germination, *Pinus wallichiana* seeds required a longer soaking period for lower GA3 concentration (100 ppm for 36 hours) than for a higher concentration (200 ppm GA3 for 24 hours) to get the comparable germination.
- Exogenous application of GA3 has been reported to be effective in breaking dormancy and substituting for the chilling requirement in the seeds of many species (Smiris *et al.*, 2006; Pipinis *et al.*, 2012).
- For enhancing the rate and percentage of germination and for breaking the dormancy, moist chilling or cold stratification has been widely used as a pre-sowing treatment (Schopmeyer, 1974; AOSA, 1992; ISTA, 1999; Wang and Berjak, 2000). This is a simple, inexpensive and effective technique for overcoming seed dormancy. Though the phenomenon is not yet fully known, the effects of moist chilling in establishing hormonal levels have been proved due to the initiation of

appropriate enzyme activity (Nikolaeva, 1977). Gibberellic acid (GA3) has been shown to promote the germination of seeds (Vogt, 1970; Krishnamurty, 1973; Chandra and Chauhan, 1976; Ghildiyal, 2003).

- Studies on both physiologically dormant *Picea glauca* and non-dormant *Picea mariana* (black spruce) seeds have shown that moist chilling was beneficial in accelerating the rate of germination (Wang, 1987). Moreover, the phenomenon of cold stratification has long been recognised in overcoming the physiological dormancy of seeds of many species. The work of Wang (1973 and 1987) on *Pinus* species and *Picea* species indicates that short periods of moist chilling of such non-dormant seeds may generally be efficacious, particularly if damage has accumulated due to natural deterioration or as a result of an imposed accelerated ageing regime. The hypothesis, therefore, is that the repair mechanism may be activated during moist chilling on non-dormant, cold temperate gymnospermous seeds, which are naturally aged or have been subjected to accelerated ageing. If this is generally applicable, cold stratification could have far-reaching implications for nursery practices, at least for such gymnosperms, in which, aged and stored seed lots are used.

Discussion

Cold stratification or chilling under moist conditions has long been recognised as a useful method of treating seeds to improve the rate and percentage of germinability (Outcall, 1991). Heydecker and Coolbear (1977) suggested that many other pre-sowing treatments increased the germination percentage and rate. Thapliyal (1986) found that under favourable conditions, the seeds of *Pinus roxburghii* germinate well (60-80%) within 7-21 days. Since most of the seeds of chirpine are non-dormant, therefore, it is often assumed that pre-conditioning treatments are unnecessary in this species. However, our results have shown that pre-chilling treatment enhances the rate of germination significantly at 20°C and 25°C. Our results are in conformation with those of Barnett (1971) who found that soaking of seeds in aerated water promotes the germination of non-dormant seeds of southern pine. There are significant differences in the germination of 15 days pre-chilled and un-chilled seeds of chirpine at 20°C and 25°C. Similar results were reported by Wang and Berjak (2000) between 14 days pre-chilled and un-chilled seeds of *Picea mariana*, which were subjected to different constant temperatures. Jones and Gosling (1994) and Jinks and Jones (1996) also reported that for the shallowly dormant seeds of Douglas fir (*Pseudotsuga menziesii*), Lodgepole pine (*Pinus contorta*) and Sitka spruce (*Picea sitchensis*), moist-chilling is a requirement to alleviate dormancy. The beneficial effects of the germination speed or rate (as a result of cold stratification) on the quality and quantity of nursery-grown seedlings were reported by Venator (1973) in Caribbean pine (*Pinus caribaea*), Mexal (1980) and Barnett and McLemore (1984) in Loblolly pine (*Pinus taeda*) and Logan and Pollard (1979) in Japanese larch (*Larix kaempferi*). Although, moist-chilling for 15 days did improve the rate and percentage of germination of the non-dormant chirpine seeds at 20°C and 25°C over 21 days, contradictory results were observed for seeds germinated at 30°C temperature. Pre-chilling treatment reduced the germination percentage in most of the provenances at 30°C, whereas in the remaining provenances, there was not much difference in the germination percentage between pre-chilled and unchilled seeds at 30°C. The effect of moist-chilling on the activation of germination in the chirpine seeds has not been previously reported and, therefore, is a new facet in understanding the benefits of short-term maintenance of seeds in a moistened condition at 3°C. The present results substantiate the beneficial effects of cold stratification on the release of dormancy and enhancing the rate of germination of chirpine seeds, which can be used in nurseries for large-scale afforestation programmes.

Conclusion

Every effort should be made to collect seed in good seed years when cones are plentiful and the quality of the seed is at its best. It is generally best to collect seed at the earliest opportunity after seed maturation as this is the time when the maximum amount of seed is available. Conifer seeds generally have physiological dormancy and this can be overcome by cold stratification (20°C-50°C). Some conifer seeds may require one month or less of cold stratification to overcome dormancy. Seed dormancy in many conifers such as *Abies alba, Abies pindrow, Abies procera, Picea smithiana, Pinus densiflora* and *Cedrus deodara* can be overcome by cold stratification and overwintering for varying periods of time from 21 to 90 days. Seed for long term storage should be dried in warm, dry air until the moisture content is between 6% - 8%. The dried seed should then be placed in airtight containers and stored in a cold store.

References

Andersson L, Milberg P. 1998. Variation in seed dormancy among mother plants, populations and years of seed collection. Seed Science Research, 8: 29-38.

AOSA, 1965.Rules for seed testing. Proc. Assoc. Offic.Seed, Anal.54(2): 1-112.

AOSA, 1992.(Association of Official Seed Analysts).Rules for testing seeds. Journal of Seed Technology; 6:1-125.

Barnett J.P.1971. Aerated water soaks stimulate germination of southern pine seeds. Res. Pap. SO67. USDA Forest Service, Southern Forest Experiment Station, New Orleans, LA, USA.

Barton, L. V. 1951. Germination of seeds of Juniperus virginiana. Boyce Thompson Institute.16: 387-393.

Baskin CC, Baskin JM. 1998. Seeds: Ecology, Biogeography, and Evolution of Dormancy andGermination. San Diego, CA: Academic Press, p. 666.

Bonner, F. T. H., Barner, Gordon and Kamra, S. K. 1974.Pinustaeda seed pretreatment. A guide to pre seed handling. FAO corporate Document Repository. Ch:8.

Borghetti, M., Vendramin, G. G., Veneziano, A. and Giannini, R. 1986. Influence of stratification on germination of Pinus leucodermis. Can. Jour. Of For. Res. 16: 867-876.

Chandra JP, Chauhan PS. 1976. Note on germination of spruce seeds with gibberellic acid. Indian Forester. 102(10):721-725.

Donelly, E. D. 1970. Persistence of hard seed in Vicia lines derived from linter specific hybridization. Crop Science. 10: 661-662.

Ellery AJ, Chapman R. 2000. Embryo and seed coat factors produce seed dormancy in cape weed (Arctotheca calendula). Australian Journal of Agricultural Research, 51: 849–854. Ghildiyal SK, Sharma CM, Khanduri VP. 2007. Improvement of germination in Chir-pine bytreatment with Hydrogen peroxide. Journal of Tropical Forest Science.19(2):113-118.

Ghildiyal, S. K., Sharma. C. M., Gairola. S. 2009. Effect of cold stratification on the germination of seeds of chirpine (Pinus roxburghii Sargent) from Indian Himalayan Region. Nature and Science. 7(8): 36-43.

Ghildiyal SK, Sharma CM, 2009. Sumeet Gairola Additive genetic variation in seedling growth and biomass of fourteen Pinus roxburghii provenances from Garhwal Himalaya. Indian Journal of Science and Technology.2(1):37-45.

Ghildiyal SK, Sharma CM, Sumeet Gairola. 2008. The Effect of Temperature on Cone Bursting, Seed Extraction and Germination in Various Provenances of Pinusroxburghiifrom Garhwal Himalaya. Southern Forests.70(1):1-5.

Ghildiyal SK, Sharma CM. 2005.Effect of seed size and temperature treatments on germination of various seed sources of Pinus wallichiana and Pinus roxburghii from Garhwal Himalaya.Indian Forester.131(1):56-65.

Ghildiyal SK, Sharma CM. 2007.Genetic parameters of cone and seed characters in Pinus roxburghii. Proceedings of the National Academy of Sciences, India 77(B),II:186-191.

Ghildiyal SK. 2003. Provenance testing in Pinus roxburghii from Western-central Himalaya. Ph.D. thesis, H.N.B. Garhwal University Srinagar Garhwal Uttaranchal, India.

Gusano MG, Gomez PM, Dicenta F. 2004. Breaking seed dormancy in almond (Pru nusdulcis(Mill.) D.A. Webb). Scientia Horticulturae, 99:363–370. Haavisto, V. F. and Winston, D. A. 1974. Germination of black spruce and jack pine seeds at0.5oC. For. Chron. 50(60): 240.

Gutterman Y. 1993. Seed Germination in Desert Plants.Adaptations of Desert Organisms. Springer-Verlag, Berlin, Heidelberg, New York, 253 pp.

Heydecker W, Coolbear P. 1977. Seed treatments for improved performance: survey and attempted prognosis. Seed Science and Technology.5: 353425.

International Seed Testing Association (ISTA). International rules for seed testing, 1999. SeedScience and Technology 27: (supplement). 1999.

Janick, J. 1974. Horticultural science with freeman and cooperation , USA, p. 188.

Jenesen, A. 1967.The influence of stratification on germination and chemical contents of immature seeds of P. abies and O. sitchensis, For.Res. Inst. Norway Rep. No. 43: 171-187.

Jinks, R. I. and Jones, S. K. 1996. The effect of seed pre-treatment and sowing dates on the nursery emergence of Sitka spruce (Picea sitchensis) seedlings. Forestry oxford 60(4):335-345.

Jones SK, Gosling PG. 1994. 'Target moisture content' prechill overcomes the dormancy of temperate conifer seeds. New Forest.8: 309-321.

Karlberg, S. 1953. Treatment of Pine and Spruce seeds to stimulate germination.Bull. R. Sch. For. Sweden. 11: 39.

Kermode AR. 2005. Role of abscisic acid in seed dormancy. Journal of Plant Growth Regulation, 24: 319–344.

Krishnamurthy HN. 1973. Gibberellins and plant growth. Wiley Eastern Limited, New Delhi 356:19114.

Logan KT, Pollard DFW.1979. Effect of seed weight and germination rate on the initial growth of Japanese larch. Bi-monthly Research Notes 35: 28-29.

Malik, A. R., Shamet, G. S. and Ali, M. 2009. Germination and seedling growth of Pinus gerardiana in nursery: Effect of stratification period and temperature. Indian Journal of Forestry. 32(2): 221-225.

Mexal JG. 1980. Growth of loblolly pine seedlings. I. Morphological variability related to day of emergence. Weyerhaeuser Co. Forest Research Technical Report 042 -2008/80/4D.

Miller WF. 1971. Duration of stratification period for Pinus elliotii.Floresta.3(2):83-85. [16] Stilinovic S, Tucovic A. Preliminary study on seed of Abies concolour Lind. from the seed stand at Avala (Yugoslavia). 95-102. Quoted from Foresty Abstracts 1971:35(10).

Mughal, A. H & R. C. Thapliyal (2006). Effects of Moisture Content & Storage Temperature on the Viability of Cedrus deodar Seeds. Seed Research, Vol. 34 (2): 182-186.

Mughal, A.H; Javeed Mugloo & Zaffar, Naseem. (2007). Effect of sowing dates on the germination and seedling growth of Pinus wallichiana and Cupressus torulosa. Indian Journal of Forestry, Vol. 30(3):295-298.

Nikolaeva, M. G. 1969. Characteristics of seed germination in gymnosperms. BotanichekiiZhurnal. 75(12) : 1648-1656.

Nikolaeva MG. 1977. Factors controlling the seed dormancy pattern. In: Khan AA, ed. The Physiology and Biochemistry of Seed Dormancy and Germination., Amsterdam, New York, Oxford: North-Holland Publishing Company,51-74.

Outcall KW.1991. Aerated stratification improves germination of Ocala sand pine seed. Tree Planters Notes.42(1):22-26.

Pawak, W. H. 1993. Germination of Alaska Cedar seed. Tree Planter's Notes. 44(1):21-24. Schopmeyer CS. 1974.Tech. Coordinator.Seeds of woody plants in the United States. USDA, Forest Service, Agriculture Handbook No. 450. Washington, DC: US Government Printing Office.

Shafiq Y, Omer M. 1969.The effect of stratification on germination of PinusbrutiaTen. Mesopotamia Journal of Agriculture.5-6: 96-99.

Shafiq Y.1980.Effect of gibberellic acid (GA3) and prechilling on germination percent of Nothofagusobliqua (Mirb.)Oerst.and N. proceraOerst. seeds. Indian Forester.106(1):27-33.

Sharma CM, Ghildiyal SK, Nautiyal DP. 2001. Plus tree selection and their seed germination in Pinusroxburghii from Garhwal Himalaya. Indian Journal of Forestry.24:48-52.

Singh, V. 1989.Seed maturity indices in Spruce.Indian Forester. 115(5): 342-346.

Singh RV. 1973.Regeneration of silver fir (Abies pindrow) forests. Proceedings of First Forest Conifers, F.R.I., Dehradun.Singh RV, Chandra JP, Sharma RK. Effect of depth of sowing on germination of spruce (Picea smithiana) seed.Indian Forester 1975;101(3):170-175.

Tamta, B. P., Verma, S. K. and Nabneta, C. 2001.Effect of moist pre-chilling on the viability and germination of Cupressus torulosa Don.Seed.Indian Forester. 127(12): 1405-1407. Thapliyal RC. 1986. A study of cone and seed in Pinus roxburghii Sarg. Journal of Tree Science 5(2):131-133.

Thompson and Morgan, 1990.Bird. R. (Editor) Growing from seed. Volume 4.
Venator CR. 1973. The relationship between seedling height and date of germination in Pinus caribaea Var. hondurensis.Turrialba. 23:473-474.
Vogt AR. 1970.Effect of gibberellic acid on germination and initial seedlings growth of Northern oak.Forest Science.16(4):453-459.
Wang BSP. 1973. Laboratory germination criteria for red pine (Pinus resinosa Ait.) seed. Proc ASOA. 63:94-101.
Wang BSP. 1987. The beneficial effects of stratification on germination of tree seeds. In: Proceedings, nurserymen's meeting, Dryden, ON, June 15-19. Toronto: Ministry of Natural Resources, 5675.
Wang BSP andBerjak P. 2000.Beneficial effects of moist chilling on the seeds of black spruce (Picea mariana (Mill.) B. S. P.). Annals of Botany.86: 29-36.
Young, J. A. and Young, C. G. 1992.Seeds of woody plants in North America, DioscoridesPress, Portland, Oregon

3

Seed Stands, Seed Production Areas Seed Orchards and Their Management

Parvez Ahmad Khan and Suhail Ahmed Wani

Faculty of Forestry, SKUAST-K, Benhama-Watlar, Ganderball, Jammu and Kashmir

Introduction

The applied aspect of tree improvement consists of the development of improved trees followed by the mass production of the improved stock. No programme will be successful until both have been achieved. Too often, the improvement of forest trees is obtained without sufficient concern about how the improved material is to be reproduced and used on an operational scale. The better trees can be regenerated through seed or by vegetative propagation. All tree improvement programmes must have seed production at some stage of their development if continued gains are to be achieved. This is even true for the programmes using vegetative propagules for large-scale operations of planting; seed is needed for the development of outstanding trees, from which, vegetative propagation can be achieved.

The organisation of extensive planting programmes need large quantities of improved seed immediately. The approach followed in their circumstances of immediate need will be somewhat different from that in which the need for seed for operational planting is still sometime in the future. Even when the results from the breeding programme may not be used until some years in the future, it is essential to set aside or establish areas for early seed production, i.e., establishment of seed stands, seed production areas and seed orchards (physiologically & genetically improved seed).

The category of 'Selected Stands in OECD's Scheme for the Control of Forest Reproductive material moving in International Trade' is similar to "a stand of trees superior to the accepted mean for the prevailing conditions and which may be treated for the production of seed", although in this case, the cultural treatment is optional rather than mandatory. The 'Registered Seed Areas' or'Selected Stands' include a mandatory thinning or thinnings to improve stand quality and ultimately, seed production for plantations.

Seed Stand (SS)

The traditional Seed Production Area (SPA) or Seed Stand, as described by e.g. Matthews(1964), Squillace (1970), Barner (1975), Keiding (1975) and Mittak (1978), is defined as ”A plus stand that is generally upgraded and opened by removal of undesirable trees and then cultured for early and abundant production of seed” (Snyder 1972). The characteristics of the traditional SPA, which may be in either indigenous forest or plantations, are:

a) It is old enough to provide reasonable assurance that it is well adapted to the site and will continue to show rapid, healthy growth and good form.
b) It is judged to be phenotypically superior to the other stands of similar age and growing in similar conditions, with which it is compared.
c) It is old enough to be bearing, or about to bear, substantial seed crops.
d) Its phenotypic quality and the seeding capacity of the residual trees are further improved by the removal of the inferior trees in the crop.
e) Prior to selection it has not usually been managed for the primary purpose of seed production.

Researchers stress the importance of providing some degree of isolation of seed production areas from foreign pollen of inferior stands or hybridising species. Complete isolation is impossible for most species and a maximum contamination of 20% of foreign pollen has been accepted for *Pinus sylvestris* in Finland (Koski 1982). As a Seed Stand reaches seed-bearing age, the differences between it and a traditional SPA become less, but the seed yield in the Seed Stand should be higher from its having been established and managed throughout for the purpose of seed production.

The Seed Stand differs from the traditional Seed Production Area in several ways:

i) Its primary purpose of seed production is already known before planting. This is advantageous because the stand can be sited to provide an appropriate combination of good access, isolation from undesirable foreign pollen and conditions for future heavy seed production. Its management can also be directed from the start towards the objective of seed production. This can include early thinning and perhaps, the application of fertilisers.
ii) On the other hand, there is inevitably an interval of time between the establishment of the stand by planting and the production of substantial quantities of seed. The SS, therefore, lacks the big advantage of the traditional SPA, which is the early availability of seed having a modest improvement in genetic quality over seeds from commercial plantations.

iii) While the judgement of the superiority of an SPA as a whole and of the individual seed trees within it is based on the phenotype (genetic superiority may be tested later through progeny tests), the superior adaptability of a given provenance used in a SS should be judged from the results of comparative provenance trials carried out on sites similar to that of the SS. For *Pinus caribaea* and *P. Oocarpa,* the best sources of information are the international trials organised by the CFI, Oxford and for *Tectona grandis* and *Gmelina arborea,* those organised by the DANIDA Forest Seed Centre.

In summary the Seed Stand may be defined as: A stand of known provenance, preferably already tested and found superior in provenance trials and of broad genetic base, which is:

1. Planted for the primary purpose of seed production,
2. Isolated to reduce pollination from outside sources and
3. Upgraded and opened by removal of undesirable trees and cultured for early and abundant seed production.

Management of Seed Stands

1. On certain sites, it is essential to take adequate measures to protect Seed Stands from fire. It is recommended that appropriate fire lines be cleared around the stands and their isolation strips, especially in grassland areas and monsoonal regions with a marked dry season. For some species, controlled burning after a certain age may be both possible and desirable within the stand. For example, in Queensland, controlled burning is prescribed for *Pinus caribaea* after 10 years (Nikles and Newton, 1980). However, this technique requires great skill.
2. Protective measures against pests and diseases may need to be carried out as required. This may include spraying against defoliators or early removal of thinnings to prevent insects from breeding in them.
3. If fencing is used, it should be regularly maintained to keep out domestic stock and wild life.
4. For tropical pines, whole crop pruning to a minimum of 2 m for access is desirable. Subsequent pruning should be confined to the periodic removal of dead or moribund branches, leaving the maximum volume of live green crown for seed production. Selective high pruning of the best trees, as carried out in some commercial plantations, should not be done in either PSSs or PCSs. Pruning of the main crop may be conveniently carried out immediately after a thinning.

No pruning of teak or Gmelina is normally necessary, but singling of basal forks from teak stumps may be necessary on some plants during the first year at the same time as weeding operations.

5. Weeding ceases at the time of canopy closure, but occasional climber cutting may be required thereafter.

Seed Production Area

Seed Production Areas (SPAs) are areas where native plants of known seed source are grown to produce seed. This can be done using a horticultural type method or as part of a mixed biodiversity planting. Greening of forest areas is encouraging and assisting land managers to establish SPAs to bolster the supply of understory species and complement the seed sourced from wild populations or Seed Collection Areas (SCAs).

The establishment of SPAs represents an innovative solution for:

- Securing much needed understory species and improving regional biodiversity.
- Ensuring a continued supply of known provenance seed.
- Providing opportunities for alternative land use and income.
- Planting multi-functional vegetation for windbreaks, soil improvement.
- Increasing the availability and genetic diversity of local native seed and.
- Providing land managers with a guaranteed purchaser and distributor of seed.

Species Selection and Planting Design for Seed Production Areas

Any planting design should allow for optimal plant growth, easy seed collection and site management and not promote excessive competition between plants or opportunities for weed dominance. It should incorporate the following features.

1. Plant more than 100 seedlings per species with seed sourced from at least 20 different widely spaced parents. This is the minimum requirement in order to maintain genetic diversity and produce high quality seed.
2. Consider the impact of shading and wind when designing an SPA. For example, larger plants on the edges of SPAs can act as a windbreak and/ or minimise internal shading.
3. Some genera (such as Grevillea and Callistemon) hybridise easily and should be physically separated or limited to one species per site.
4. Maximise accessibility for management tasks such as watering, mowing, pruning and harvesting by locating the site in close proximity to facilities.

5. Match species selection with local climatic and soil conditions of the site. For example, factor in frost hollows and sites that can become waterlogged.
6. Give every species an equal chance of pollination. Block plantings of similar species are much more effective for cross-pollination from neighbouring plants than mixing up species or long lines of the one species.
7. Try to locate the site close to natural bushes to assist with pollination, especially from insects. Beware, however, as non-local seed production could escape into the bush or hybridise.

Site Establishment and Maintenance

1. Fence site off from stock and maintain, where necessary.
2. Maintain firebreaks and ensure vegetative growth remains low.
3. Soil preparation prior to planting should include ripping (up to 6 months in advance) and preferably mounding, particularly when there is poor drainage or waterlogging. Mounds should rise approximately 30cm above ground level.
4. Inter-row spacing widths for each species will vary, depending on plant height, form and site aspect. Plant density needs to factor the type and size of the plant when fully grown and can be planted so foliage is touching adult stage. Close plantings will maximise space and consolidate management, infrastructure and harvesting.
5. Plastic woven weed mats or other similar mats can be laid to greatly improve weed control. Mulch can also be used with or without weed mats, but must not be contaminated with weed seed.
6. Smaller plants will benefit from a reliable watering system, such as the dripper hose or micro-drip design. Regular watering or irrigation checks include replacing mulch or weed mats and inspection of irrigation connections, fittings and drippers.
7. Label each block or row using stakes, aluminium tags or other permanentmarkers to define species and provenance code.
8. Minimise predation damage by rabbits, kangaroos, insects, birds etc. by using a combination of deterrents, such as fences, tree guards and bird netting.
9. Periodically inspect the site for insect attack, especially when seed has set. Wattle and pea seed is particularly prone to insect and bird attack.
10. Prune plants to limit plant height and encourage large, open crowns to facilitate fruiting and seed production.

11. Ensure plants are healthy and not suffering from dieback, disease or nutrient deficiency.
12. Replace losses and keep in mind that some smaller plants may need replacing after 5 years to maintain yields.
13. Weed control is critical to the success of any planting, particularly seed production areas. Weed competition for light, space, water and nutrients will severely limit the growth and survival of plants as well as access.

Seed Orchard

A seed orchard is defined as "a plantation of selected clones or progenies which is isolated or managed to avoid or reduce pollination from outside sources, and managed to produce frequent, abundant and easily harvested crops of seed" (Fellberg and Soegaard 1975).

A seed orchard is an area where seeds are mass produced to obtain the greatest genetic gain as quickly and inexpensively as possible (Zobel and Talbert, 1984).

Types of Seed Orchards

The seed orchards can be classified variously

On the basis of the method of production:

On this basis, seed orchards are of two types

a. Seedling seed orchards

These are raised from seedlings obtained from seeds of plus trees followed by roughing which will remove the poorest trees and allow the best trees to remain for seed production.

b. Clonal or vegetative seed orchard

These are composed of vegetatively propagated trees that are raised from cuttings, which can yield seeds comparatively easily. They are also raised by grafts or grafting clones in the form of scions or buds of plus trees on the stock of seedlings raised at proper spacing in advance.

Comparison Between Clonal & Seedling Seed Orchards

S.No.	Clonal Seed Orchard	Seedling seed orchard
1	Clonal Seed orchards are raised through vegetative propagation thereby maintaining the genetic identity of the material.	Seedling seed orchards are raised through seeds, thereby loosing genetic identity through segregation.

2	The seeds obtained will be of high genetic value and good combining ability of known individual tree.	The seeds obtained may be of good genetic value.
3	Selfing is reduced by avoiding the planting of the same ramet nearby.	Selfing cannot be ruled out as pollination may take place within the family.
4	The grafted plants start flowering earlier, thereby reducing the cost of production and increasing seed availability earlier	The seedlings take more time for flowering, thereby increasing the cost of production and delay seed supply.
5	Since flowering occurs at lower height, control cross-pollination can be conducted.	The seedlings flower after completing the vegetative phase Hence their height is more at the time of flowering, which hampers the controlled pollination.
6	Because of low height of the ramet, it is easier to harvest fruits.	It is difficult because of more height of the ramet.
7	20 to 40 clones are used in an orchard (per ha.).	100 or more parental trees are used (per ha)
8	Cost of establishment is higher.	Cost of establishment is lower.

Clonal seed orchards can be of following types:

i. First generation seed orchard; which consists of plus trees in which genetic make up of selected trees are not tested.

ii. Second generation seed orchard; which consists of tested clones.

iii. Third or subsequent generation seed orchard, which has clones of a desirable selection of the best trees in the best families from the progeny trials of plus trees. The second, third or more advanced generation orchards represent the number of cycles of improvement.

Based on the Nature and Derivation of Original Trees, the Seed orchards are Divided into Following Four Types

1. Plus Tree Seed Orchards

These are orchards composed of clones of plus trees of the same species from different populations within a provenance zone.

2. Provenance Crossing Seed Orchards

These are orchards composed of a relatively large number of clones of the same species belonging to at least two provenances. These are based on the reciprocal crossing of plus tree clones of different provenances, each with a high utility value.

3. Seed Orchards for Inter Specific Hybrid Seeds

These are orchards intended for intercrossing of clones representing different species that are able to cross. Hybrid vigour can be exploited by this method.

4. Elite Seed Orchards

These are orchards composed of only controlled and genotypically good clones derived from different plus trees and provenance seed orchards. The clonal genotype and its effects of intercrossing are ascertained by progeny trials (*Source:* Kumar, 1994).

Design of Seed Orchards

These include

1. Pure Rows
2. Chessboard
3. Completely Random
4. Randomised Complete Block
5. Fixed Block
6. Rotating Block
7. Reversed Block
8. Unbalanced Incomplete Block
9. Balanced Incomplete Block

Management of Seed Orchards

The management of seed orchards should be directed towards the early establishment and healthy development of grafted trees and promotion of sustained and regular fruiting. The costs incurred in achieving these objectives should be kept as low as possible. The following management practices are applied in seed orchards:

1. Ground and ground water management: Weeds and cover crops should be kept under strict control within 0.5 metre radius of each plant until it is well-established and growing vigorously. Land that is completely cultivated should be harrowed or treated with suitable weedicide as an alternative. Grass cover should be removed regularly, but when the trees approach flowering age, it may be advantageous to encourage water stress on the site through transpiration losses by allowing the ground cover to grow tall during the period of flower formation.
2. Fertilisation: Fertilisers containing N, P or K are commonly used and have shown both positive and negative effects on flower and cone production in many species. It is not possible to provide reliable rules and prescriptions for general guidance for their use. It is clear from various studies that a more critical approach to the effects of fertilisers and better records of the basic site parameters are needed. In particular, more information is needed on the effects of fertilisers on pollen

production, female flower and cone production and seed yields in terms of viable seed, rather than yield. A negative effect of fertilisation was observed in a Larix decidua orchard, which resulted in the reduction of female flowers.

In the case of loblolly pine, the following standard has been outlined by Davy (1981).

Calcium	400 kg/ha
Magnesium	50
Potassium	80
Phosphorus	40

The timing of fertiliser application is critical if satisfactory results are to be obtained. It should be applied just before the initiation of floral buds, if immediate and increased flowering is to be obtained. Fertilisers also help the trees to remain healthy and grow to a large size, resulting in more reproductive bud locations.

3. Irrigation: It is one of the important management practices in seed orchards. Irrigation should be done when the soil is dry. Irrigation from overhead lines can be used to reduce frost damage to Douglas fir flowers. Apart from fire protection, more rapid development of trees and better sod cover, irrigation now appears to be a good investment for some species. In loblolly pine orchards, irrigation as well as fertilisation were responsible for an approximate 30% increase in seed production over one that was only fertilised and for a 100% increase over the area that received neither irrigation nor fertiliser.
4. Thinning: The initial spacing of orchard trees should be sufficiently wide so as to encourage the full development of the crown for a period long enough to ensure that cones or fruits will be obtained for several years before thinning is required. Thus, narrower spacing can be used for species which flower precociously than for species which are slower to start flowering. In an orchard, the tree crowns should always be kept isolated to admit full sunlight and to permit the easy movement of cone or seed harvesting equipment.
5. Pruning: The trees in orchards should be so managed as to make seed collection easy and convenient. Thus, the trees should be treated in a way which would encourage the development of short, wide and bushy habit, except for species where ground collection is resorted to (oak, beech etc.). However, severe pruning may reduce cone production as found in Norway spruce and Scots pine. A standard practice in South

African pine orchard management is to artificially band both the main stems and branches, a method shown to be more effective than pruning. In removing the top of conifers, many future potential flowering points are lost and the proportion of male to female flowers is increased within a zone of crown where selfing is likely to occur. This has a harmful effect on seed yield and quality.

Conclusion

The purpose behind the establishment of seed stands, seed production areas and seed orchards is that they serve as a reliable source of seed of known origin. These help in the production of hardy trees in respect of drought, diseases, frost etc., which is essential for raising nursery stock, so that the afforestation programmes will be highly productive. These serve the purpose of procuring and providing sufficient quantities of germplasm of high quality to satisfy national and international demands. These can also be used for the conservation of important forestry species which are in danger of extinction and/or rapid genetic erosion.

References

Barner, H. 1975. Identification of sources for procurement of forest reproductive material. Report on FA0/DANIDA training course on Forest Seed Collection and Handling, Vol. II pp. 42-64. FAO, Rome.

Davey, C.B. 1981. Seed orchard soil management. North Carolina State University-Industry corporative tree improvement programme. School of forestry resources, Raleigh: 90-95. Fellberg, L. and Soegard, B. 1975. Historical review of seed orchards, forestry Comm. Bull No. 54.

Keiding, H. 1975. Seed stands. Report on FAO/DANIDA training course on Forest Seed Collection and Handling, Vo1.2 pp. 192-211. FA0, Rome.

Koski, V. 1982. Quantified standards for regional clonal seed orchards. Forest Genetics Resources Information. FAO No. 11.

Kumar, V. 1994. Nursery and Plantation Practices in Forestry: 494.

Matthews, J.D. 1964 Seed production and seed certification. Unasylva Vol. 18, Nos.2-3, pp, 104-108, FAO, Rome.

Mittak, W.L. 1978 Manual 2 para la recoleccion de semillas forestales. INAF0R-BANSEF0R-FA0/TCP. Instituto Nacional Forestal, Guatemala.

Nikles, D.G. and Newton, R.S. 1980. Inventory and use of "provenance resource stands" of Pinus caribaea var. hondurensis in Queensland. Paper for IUFRO joint symposium and workshop, Brazil.

Snyder, E.B. 1972. Glossary for Forest Tree Improvement Workers. USDA, Forest Service

Squillace, A.E. 1970. Development and action programmes for forest tree improvement. Unasylva 24,Nos.97-98,p.65.

Zobel, B. and Talbert, J.1984. Applied Forest Tree Improvement: 167-216

4

Seeds Requirement in Nursery and Plantation Area: Calculations Thereof

G.M. Bhat and M.A. Islam

Faculty of Forestry Benhama, Ganderbal-191201, Jammu and Kashmir, India

Part A

Counting of Seeds in a Seed Lot

Counting of seeds is essential in order to estimate the number of seeds required for planting in a nursery or direct sowing or where seeds are valuable and are sold or purchased on number basis. However, in practice, it is difficult to count the seeds in the entire lot. Alternately, the whole lot is weighed accurately. From the whole lot, 5-6 small seed samples of 100 g each are randomly selected. The number of seeds in these samples are counted. The average of these samples is then applied to the whole seed lot in order to calculate the number of seeds.

Example: Count the number of seeds of Deodar lot weighing 18 kg.

First 5 small sample of 100 g each are randomly selected from the 18 kg lot. The seeds are counted in these samples. Let the number of seeds in the samples be as under:

- 1st sample (100g) - 50 seeds,
- 2nd sample -53 seeds,
- 3rd sample - 49 seeds,
- 4th sample -52 seeds and
- 5th sample -53 seeds.
- The average number of seeds in the sample = 51.4 seeds/ 100 grams.
- This means 514 seeds / kg. Therefore, in 8 kg, there will be 4112 seeds.
- To get more accurate results, the number of samples can be increased, depending upon the time and labour in hand.

Quantity of Seeds Required for a Given Nursery Area

A rough estimate of the quantity of seeds that will be required for sowing in a specified bed can be calculated with the following formula:

$$W=\frac{A\times D}{P\times N}\times 100$$

- W = weight of seeds required (g)
- A = Area of nursery bed (m^2)
- D = number of plants required per m^2
- P = plant percent of species
- N = number of seeds per gram

Example: Calculate the quantity of Deodar seeds required for a nursery bed of 3 x 1.5 m, which are to be sown at a distance of 25 x 25 cm. The viability of the seeds is 70 % and the number of seeds per gram is 10.

Given: Area of nursery bed=3 x 1.5 = 4.5 m^2

$$\text{Number of plant requried} / m^2 = \frac{\text{Area}}{\text{Spacing}}$$

$$\frac{1}{0.25*0.25} = 16\,\text{plants} / m^2$$

- Plant percent = 70
- Number of seeds per gram = 10

Putting these values in formula we get : $W=\frac{4.5\times 16}{70\times 10}\times 100$

$= 10.28$

For the broadcast method of sowing in nurseries, the quantity of seeds calculated is multiplied by six and for the drill sowing method, the quantity is doubled. Generally, broadcast sowing is done for minute seeds, such as eucalyptus and poplar.

Quantity of Seeds Required for a Given Plantation Area

Kuller *et al.* (1991) gave the method of calculation for determining the quantity of seeds required for a given plantation. The method of calculation is as follows:

$$\text{Quantity of seeds (kg)} = 125\times\frac{n\times A}{P\times W}+C$$

- N = number of seedlings to be planted per ha.
- A = area (ha) of plantation
- P = plant percent

- W = number of seeds per kg
- C = extra quantity to cover germination failure of some seeds.

Example: Plantation of apricot is to be done in an area of 6 hac. with a spacing of 5 x 5 m. The plant percent is 60 and the number of seeds per gram is 6. About 300 extra seeds are required to cover germination as well as plant failure. Calculate the quantity of seeds required.

$$\text{Quantity of seeds (kg)} = 125 \times \frac{400 \times 6}{6 \times 6000} + 300$$
$$= 308.08 \text{ kgs of seed}$$

However, the quantity of seeds calculated from the above formula will, in most cases, not be the actual seed required for the given plantation because it does not taken into account the number of plantable seedlings per kilogram of seed. The following formula overcomes this drawback:

$$\text{Weight of seed needed (kg)} = \frac{n \times A}{S \times P_s}$$

- N = number of seedlings to be planted per ha
- A = area (ha) of plantation
- P_s = estimated number of plantable seedling per kilogram of seeds.
- Information regarding n, A and S in above formula is readily available. However plantable seedlings per kilogram need to be worked out by following formula

P_s = N X G X P_a X (1-M)

- N = number of seeds/kg
- G = germination percent, (as decimal)
- P_a = purity of seed lot, (as decimal)
- M = seedling mortality during nursery stage (as decimal

Example: If a plantation of 7 ha of Sal is to be done with a spacing of 2 X 5m and plant per cent of Shorea is 70 percent. After removal of the 15g of impurity from 3 kg seed lot number of seeds were 1185.Germination revealed that the lot has 98 per cent germination and in nursery it performed cent per cent. Calculate the actual amount of seeds required.

Solution:

- Area to be planted=7 ha
- Spacing = 2 X 5 m
  - $\text{No. of plants required / hac} = \frac{100 \times 100}{2 \times 5} = 1000 \text{ seedling}$
  - = 1000 seedling will be planted

- Survival after planting = 70 (decimal fraction 0.70)
- Weight of seed lot = 3kg
- Impurities = 15 g
- So the weight of pure seed = 3kg-15 g = 2.985 kg
- Thus the purity of seed lot is = 99.5% or 0.995(decimal fraction)
- Germination percent = 98% or 0.98 (as decimal fraction)
- Since 2 kg 985 g contains = 1185 seeds.
- 1 kg will have = 396.98 seeds.
- First we calculate plantable size seedling (P_s)
- **Ps = N X G Pa X (1-m)**

  = 396.98 X 0.98 X 0.995(1-0)

  = 387.095

Substituting the values in the given formula to calculate the actual quantity of seeds required:

$$\text{Weight of seed needed(kg)} = \frac{n \times A}{S \times P_s}$$

$$\frac{1000 \times 7}{0.70 \times 387.095} = 25.83\,\text{kg}$$

Yet another formula (though a modification of the above) has been put forward to calculate the quantity of seeds required for plantation .The formula is:

$$\text{Quantity of seed (Kg)} = \frac{A \times D}{S \times N \times G \times M}$$

Where

- A = area (ha)to be planted
- D = destiny of plant/ha
- S = survival % of plantation (as decimal)
- N = number of seeds/kg
- G = germination percent, (as decimal)
- M = mortality per cent in nursery (in decimal)
- Calculation is similar to that of above formula.

5

Plant Growth Regulators and Their Applications in Forestry

Zaffar M. Dar[1], A.H. Mughal[1], Malik Asif[1], Amjad Masood[2] and M.A. Malik[1]

[1]*Division of Basic Sciences and Humanities, FOA, Wadura-SKUAST, Jammu and Kashmir, India*

[2]*Division of Agronomy, FOA, Wadura- SKUAST-K, Jammu and Kashmir, India*

Introduction

Plant growth regulators are defined as small, simple chemicals produced naturally by plants to regulate their growth and development and are essential for regulating their own growth. They act by controlling or modifying plant growth processes, such as formation of leaves and flowers and elongation of stems.

In modern forest nursery raising, the benefits of extending the use of plant hormones to regulate the growth of economically important plants have been established. There are two possible goals in inducing trees to form floral buds, in particular, female buds. The first is to achieve precocity. Precocious flowering in normally immature trees can provide advantages in breeding and seed production because it overcomes a substantial seed production delay intrinsic to late-maturing conifers. The second reason is the phase change from juvenile growth to mature growth. The induction of either precocity or increased male flower and female cone formation are generally the result of applied treatments. A few studies have focused on the intrinsic physiological differences between high and low producing clones. The traditional methods for improving seed yield involve treatments to manipulate the physiological conditions of the parent trees, which in turn, enhances flowering. Flower enhancement is achieved by either physically stressing the trees, altering the trees' nutritional status by applying various inorganic fertilisers (Sweet and Hong, 1978) or applying growth regulating substances such as gibberellins (GAs) (Owens, 1991). Physical treatments are either applied to the shoot (top working, shoot training, girdling, wire girdling, scoring) or the root (pruning, trickle irrigation, fertigation, confinement). Some treatments are combinations of these general categories of tree management to enhance the effects of each treatment (Ross and Bower, 1991). The most notable method for cone induction is the application of plant growth regulators due to its efficiency and practice in application. The five major types of plant hormones include gibberellins, cytokinins, auxins, abscisic acid and ethylene. This review focuses on plant hormones or growth regulators and their application in flower enhancement in forest species.

Characteristics of Plant Growth Regulators

Plant Growth Regulators can be of a diverse chemical composition, such as gases (ethylene), terpenes (gibberellic acid) or carotenoid derivates (abscisic acid). They are also referred to as plant growth substances, phytohormones or plant hormones. Based on their action, they are broadly classified as follows:

- Plant growth promoters: They promote cell division, cell enlargement, flowering, fruiting and seed formation. Examples are auxins, gibberellins and cytokinins.
- Plant growth inhibitors: These chemicals inhibit growth and promote dormancy and abscission in plants. Abscisic acid is an example of the same.

Role of Different Growth Hormones

Auxins

Auxins were the first growth hormone to be discovered through the observations of Charles Darwin and his son, Francis Darwin. Phototropism formed the basis for the discovery of auxins. The term 'auxin' is a Greek word which means 'to grow'. Compounds are generally considered auxins if they have functions similar to indoleacetic acid (IAA), the first auxin isolated from plants. The typical function is to induce cell elongation in stems.

The response of coniferous trees to auxin induction vary by species. Auxins such as naphthaleneacetic acid (NAA) are usually applied with GAs in order to enhance the effects of the latter (Pharis *et al.*, 1980). In these combinations, low NAA concentrations favoured female flower development in Douglas fir, whereas high NAA concentrations stimulated male flower formation. The application of NAA alone decreased female flower formation in lodgepole pine and Chinese red pine (Sheng and Wang, 1990). However, the application of NAA alone enhanced male flowering in Chinese red pine (Sheng and Wang, 1990). In combined applications, the addition of NAA enhanced the GA4/7 effect in Loblolly pine, but diminished it in Slash pine and Longleaf pine during male flower bud induction (Hare, 1984).

Adventitious root formation is often a limiting step for in -vitro vegetative plant propagation programmes. Like many species, adventitious roots are difficult to form in woody plants. The results of the present investigation indicated that it is effective to root by stem cuttings of hybrid aspen with the application of auxins. Auxins are known to play a significant role in stimulating AR from stem cuttings of tree species (Tchoundjeu *et al.*, 2004). In addition to enhancing the rate of AR development, auxin application has been found to increase the number of roots initiated per rooted cutting in a variety of species. Among the growth hormones, NAA at 0.54 mM have been found superior in terms of rooting rate and number of rootings and hence, it was recommended for rooting in hybrid aspen. This is in line with the fact that a particular type of auxin is effective in enhancing rooting in a particular species (Puri *et a*l., 1988). A similar trend was also recently recorded for hormonal applications at different concentrations by Tiwari and Das (2010) who reported that NAA are known to promote the expansion of roots in cutting, thereby increasing the survival rate. It is worth noting that higher concentrations of auxin did not substantially produce better results of rooting.

Gibberellins

It is the component responsible for the 'bakane' disease of rice seedlings. The disease was caused by the fungal pathogen *Gibberella fujikuroi*. E. Kurosawa treated uninfected rice seedlings with sterile filtrates of the fungus and reported the appearance of disease symptoms. Finally, the active substance causing the disease was identified as gibberellic acid. There exist more than 110 gibberellins, obtained from a variety of organisms from fungi to higher plants. They are allacidic and are denoted as follows – GA_1, GA_2, GA_3 etc. GA_3 (Gibberellic acid) is the most noteworthy since it was the first to be discovered and is the most studied.

The less polar GAs, such as gibberellins GA4, GA7 and GA9, were the most effective ones in flower induction in *Pinaceae* (Pharis, 1991). The preference of different GAs in different plant families depends not only on their stability, but also on the turnover rate of GA metabolism in plant tissue. The application of such GAs alone or in combination with other methods stimulated cone induction in Douglas fir (Pharis, 1991), lodgepole pine (Longman, 1983), eastern white pine (Pijut, 2002) etc. Effective stimulation of cone initiation and differentiation by GAs has been largely successful in conifers.

Eastern black walnut (*Juglans nigra L.)* is used as a rootstock for the Persian walnut *(Juglans regia L.)* in some parts of the world and also has an important role in forestry and the wood industry. Due to deep physiological dormancy, the seed often shows an inconsistent or low germination percentage, making establishment difficult. Results have shown that the germination rate with chilling treatment was zero, whereas, the highest percentage of seed germination (69.27%) was recorded with the combined treatment o f chilling and GA_3(400 ppm). Also, this treatment showed significant differences for morphological, physiological and biochemical parameters compared to other treatments. It was found that the application of the combined treatment of chilling stratification and$_3$ GA was effective in increasing the seed germination percentage and rate as well as improving the growth parameters of Eastern black walnut seedlings. The effect of GA3 and stratification on enhancing growth could be attributed to the solubility of fats and sugars due to stratification as well as the increase in the synthesising of gibberellins. In addition, the improving effect of GA3 and stratification on seed germination may reflect on the enhancement of the shoot parameters.

Cytokinins

Cytokinins are compounds are adenine derivatives and promote cell division. Cytokinins have been shown to play a significant role in tree development in a wide variety of pinaceous trees. Exogenously applied cytokinins increase bud initiation and also control bud release. Exogenous applications of gibberellins that induce bud initiation have been accompanied by an increase in endogenous cytokinin levels (Pilate *et al.*, 1990). It is known that spring cytokinin levels increase in the root and that these compounds are transported via the xylem to shoots where they influence development. Morris *et al.*, (1990) reported that all buds in Douglas fir have three major cytokinins. In female and vegetative buds, the zeatin-type cytokinin was relatively higher, while in male flower buds, the concentration of isopentenyladenosine was higher with low levels of zeatin-type cytokinins. Buds of radiata pine contained different cytokinins and varying patterns were observed in developing and mature buds. The implication is that bud development is, in some way, regulated by cytokinins (Zhang *et al.*, 2003). Cytokinins have been used alone or in combination with other PGRs to induce cones. Benzlaminopurine (BAP), a synthetic cytokinin, enhanced GA-induced flowering in Douglas and Sitka spruce. Recently, significant progress was made in cytokinin-linked cone induction. When BAP was applied to Japanese black pine and Japanese red pine at specific developmental stages, female flower formation was greatly enhanced through the conversion of male flower buds into female ones. The most effective treatment was a combination of BAP with girdling (Wakushima, 2004). This method shows great potential since cone induction rates by BAP were three to tenfold higher than those by GAs in the same species.

Abscisic Acid

Abscisic acid (ABA) was initially discovered to play a role in leaf abscission in the autumn. ABA has been found to be a key factor in regulating many physiological responses, including transpiration, stress response, germination of seeds and embryogenesis. ABA influences plant growth and development, but often by interactions that involve other plant hormone pathways. For optimal growth, plants require an optimal level of ABA. Plant growth suffers when ABA levels are too low or too high. ABA is the major player in bolstering the adaptation of the plants to stress. Drought treatments can enhance flowering and seed yields in both conifers and angiosperm trees, especially if they are applied during the period of shoot elongation. It has been known since the early 1980s that drought-treated conifers send ABA from their roots into their shoots within 10 hours of such a treatment. Some studies show that bud differentiation may well involve, in part, a response to increased ABA metabolism. Other studies imply an interaction between cytokinins and ABA in these processes. Girdling and pruning are commonly used methods for enhancing flowering in conifers. These physical treatments may influence endogenous ABA concentrations in the stressed trees. It was found that flower induction by GAs also involved increased endogenous ABA concentrations. In Douglas fir, exogenous GA4/7 enhanced flowering, while doubling ABA concentrations in GA-treated trees compared to controls (Pilate *et al.*, 1990).

Application of PGRs

Stimulation in female or male flower bud differentiation not only relies on the types of treatments, but also on the developmental stages of the buds at the time of treatment. Different results may arise from the same treatment applied at different times. To determine the correct application time, it is first necessary to establish which developmental stages are most sensitive to PGR treatments. The responses of plants to induction treatments vary by species and PGRs. GAs are used before or during the initiation stage of floral buds in order to enhance flowering, whereas cytokinins, such as BAP, are used at the floral differentiation stage for converting the sexual expression of male buds to female buds. The application of BAP to plants before or during the initiation stage did not influence sexuality, but only enhanced vegetative growth. If BAP was applied too late, i.e., after the differentiation stage, there was no effect at all. Thus, timing is more important than the frequency of PGR application.

PGR Quality and Quantity in Applications

The quality of the chemicals can make a difference. Of all the synthetically produced chemicals (e.g. NAA, BAP and various GAs), gibberellins are the most notoriously variable. A mixture of GA_4 to GA_7 is commonly used for cone induction. The ratios of GA_1 to GA_4 vary according to different supplement sources. Trace amounts of other GAs, such as GA and GA may exist in floral induction in Sitka spruce. The total amount of applied PGRs must be determined in relation to the size of the trees (i.e., larger tree, more PGR), after which, concentrations must be calculated in relation to the application method, such as stem injection or foliar spray.

PGR Application Methodology

The application methods of PGRs include foliar spray, topping, paste, bud injection, stem injection etc. Applications of PGRs are usually combined with other methods such as girdling, pruning, drought and/or fertilisation.

References

Hare RC. 1984. Application method and timing of gibberellin A4/7 treatments for increasing pollen conebud production in southern pines. Can J For Res 14: 128-131

Longman KA. 1983. Effects of gibberellin, clone and environment on cone initiation, shoot growth and branching in Pinuscontorta. Ann Bot 50: 247–257.

Morris JW, Doumas P, Morris RO, Zaerr JB. 1990. Cytokinins in vegetative and reproductive buds of Pseudotsugamenziesii. Plant Physiol 93: 67-71.

Owens JN. 1991. Flowering and seed set. In Physioloy of trees.Edited by AS Raghavendra, Wiley, New York. pp. 247-271.

Pharis RP, Ross SD, McMullan EE. 1980. Promotion offlowering in the Pinaceae by gibberellins. III. Seedlings of Douglas-fir. Physiol Plant 50: 119–126

Pharis RP. 1991. Physiology of gibberellins in relation to floral initiation and early floral differentiation. In Symp.on 50th Anniversary Meeting on Isolation of Gibberellins. Eds. N. Takahashi, B.V. Phinney and J. MacMillan. Springer Verlag, Heidelberg, pp.166-178

Pijut PM. 2002. Eastern white pine flowering in response to spray application of gibberellin A4/7 or procone. NJAF 19: 68-72

Pilate G, Sotta B, Maldiney R, Bonnet-Masimbert m, Miginiac E. 1991. Endogenous hormones in Douglas-fir trees induced to flower by gibberellin A 4/7 treatment. Plant PhysiolBiochem 28: 359-366.

Puri, S.; Shamet, GS. 1988. Rooting of stem cuttings of some social forestry trees. Int. TreeCrop J., 5, 63-69.

Ross SD, Bower RC. 1991. Promotion of seed production in Douglas-fir grafts by girdling + gibberellin A4/7 stem injection, and effect of retreatment. New Forests 5: 23-34

Sheng C, Wang S. 1990. Effect of applied growth regulators and cultural treatments on flowering and shoot growth of Pinustabulaeformis. Can J For Res 20: 679-685

Sweet GB, Hong SO. 1978. The role of nitrogen in relation to cone production in Pinusradiate. NZJ ForSci 8: 225-238

Tchoundjeu, Z.; Ngo Mpeck, M.L.; Asaah, E.; Amougou, A. 2004.The role of vegetative propagation in the domestication of Pausinystaliajohimbe (K. Schum), a highly threatened medicinal species of West and Central Africa.For. Ecol. Manage. 188, 175–183

Tiwari, R.K.S.; Das, K. 2010.Effect of stem cuttings and hormonal pre-treatment on propagation of Embeliatsjeriam and Caesalpiniabonduc, two important medicinal plant species.J. Med. Plants Res. 4(15), 1577- 1583.

Wakushima S. 2004. Promotion of Female Strobili Flowering and Seed Production in two Japanese Pine Species by 6-Benzylaminopurine (BAP) Paste Application in a Field Seed Orchard. J Plant Grow Regul 23: 135-145

Zhang H, Horgan KJ, Reynolds PHS, Jameson PE. 2003. Cytokinins and bud morphology inPinus radiate. Physiol Plant 117: 264-269

6

Nursery Raising of Conifers Under Temperate Conditions of Kashmir Valley

J.A. Mughloo, K.N. Qaiser, Mehraj u Din Dar and Rameez Raja

Faculty of Forestry Benhama Ganderbal 191 121, SKUAST-K, Jammu and Kashmir, India

Introduction

Forest restoration is a complex process that requires many steps to ensure successful forest establishment. These steps include choosing suitable tree species and provenances, applying nursery cultural practices to produce quality seedlings and making site modifications to improve the physical environment of the restoration site (Grassnickle, 2000). Conifers are the most important tree species of the western Himalayas and J&K. There are about 16 species spread over 9 genera in 3 families. Among these families, *Pinaceae,* with seven species in 4 genera, is the most dominant, while *Taxodiaceae,* with 2 species in 2 genera, is the least represented. Out of the total taxa, 7 species belonging to 5 genera are exotic and exist in cultivation only. They occupy an area of about 8.27 lakh hectares and constitute an important floristic component of evergreen forests by virtue of their multi- dimensional ecological and socio-economic values. Nowadays, conifers are not only used for afforestation and reforestation purposes, but are also being extensively used for ornamental purposes because of their beauty and evergreen nature. As such, the demand of conifer seedlings is on the rise. Besides, conifer nursery raising has a tremendous scope for employment generation as a huge market for seedlings is available throughout the valley, both in the private and government sector. Raising seedlings in nurseries can serve the purpose of reforestation and also provide enough plant material for planting outside conventional forests. Raising seedlings on modern scientific lines will help in achieving maximum germination and survival, besides obtaining quality seedlings in a shorter period, thus minimising the costs of nursery raising.

Following conifers are generally raised in nursery

Deodar (*Cedrus deodara*)

Cupress (*Cupressus torulosa*)

Kail (*Pinus wallichiana*)

Halepensis (*Pinus halepensis*)

Fir (*Abies pindrow*)

Spruce (*Picea smithiana*)

Steps Involved in Quality Planting Material

Seed collection

Seed collection can be done from standing trees by climbing them or from felled trees.

Seed collection should be confined to good seed years when seed quality is at its best and collection costs are lowest.

Cone-bearing seeds should be harvested from trees in the second fortnight of October. Cones should be collected before they break open.

Seeds should be collected from healthy and middle- aged trees of good form and not from over- matured trees or malformed ones.

Cones should be sun- dried after collection.

Nursery Site

A level or nearly level, well- drained site suitable for the production of seedlings should be chosen

The site should be in or near the area where the plantation is carried out.

Deep, fresh, light and sandy loam soil meets most of the nursery requirements

The site should have irrigation or perennial water supply

The site should be well- connected with roads for easy transport of plants and other nursery material

Soil Media

Soil media should be mixture of soil sand and well decomposed FYM in the ratio of 2:1:1, pH should vary from 5.5-7.5

Containers

Containers in the form of polybags or root trainers are used.

Various types of poly bags are available in the market and any one of them can serve the purpose, provided it is of suitable size

The optimum size and thickness of polybags should be 40x25 cm with 200 gauge thickness.

Root trainers are specialised containers devised for nursery raising and are nowadays used because of the number of advantages they offer over polybags. They are available in different capacities and root trainer of 150 cc is ideal for raising seedlings, Besides they can be reused and are easy to handle.

Seed Pre-treatment

As most of the conifers have mild internal dormancy and hard seed coat, they need some pre- sowing treatments to break the dormancy and overcome the hard seed coat

Different treatments are given to different species of conifers.

Pine and spruce seeds are soaked in water for about 10 days at room temperature.

Cupressus torulosa and *Pinus halepensis* seeds should bemoist chilled for 7 days

There are different methods of seed pre-treatment, depending on the type of dormancy. Very often, a method for a certain type can also work for another type, so as to hit 'two birds with one stone'

a) **Physical Methods:** For seeds with exogenous dormancy (impermeable, hard seed coat), common techniques of pre-treatment include scarification, which involves rubbing the seed against a rough surface (file, sand paper, concrete mixers, seed scarifiers or any other abrasive material) or nipping/ nicking, so as to make a very small 'hole' sufficient to allow water to enter. This method is more applicable to seeds with seed coat dormancy, but not for resinous or pulpy fruits

b) **Hot/Cold water treatment:** Such treatments combine the effects of softening hard seed coats and leaching out chemical inhibitors. Some seeds that have little resistance to germination may respond well to soaking for 24 hours in water at ambient temperature. A more effective treatment, especially in hot climates, is alternate wetting and drying of seeds, such as one day's soaking followed by three to four days of drying. In hot water treatment, seeds are usually placed in boiling water which is immediately removed from the heat source and left to cool gradually. The seeds remain in the water for about 12 hours. The ratio of water to seeds can be determined by experimentation and may vary considerably according to the species- 2-3 times, 4-5 times and 5-10 times. By and large, the volume of water should be much higher. (Refer part VII number 15.3.3(c),UGIS of 2015 NB) Care should be taken not to damage the embryo through excessive heat, especially if the seed coat is permeable.

c) **Chemical treatment/Acid treatment:** Concentrated H SO (95%) is commonly used to treat dormancy, especially for seeds that have been kept in storage for a long time.

Materials: Acid (SG = 1.84, 95% purity), acid resistant containers, screens for handling draining & washing seeds, abundant supply of water for rinsing.

Procedures: Allow the seeds to come to room temperature. Mix the seeds thoroughly. Immerse the seeds in the acid (duration based on experiment). Remove the seeds from acid and rinse with abundant water for 5 to 10 min. Spread the seeds on a mat for drying. Care is required in handling acid as its excessive application may damage the seeds.

d) **Cold stratification:** This method is usually applied to treat physiological dormancy, mainly for temperate species, but is also effective for some tropical highland species. Stratification refers to the method of placing seeds in layers, alternating with layers of a moisture- retaining medium, such as sand/peat and keeping them at a cool temperature for a certain period, which is commonly between 20-60 days, but varies considerably from species to species. Pre-treatment of seeds includes soaking in cold water/moisture medium at a temperature of 3-5°C.

Seed Sowing

The containers to be used should be filled up to 90% of their capacity, in case polybags are used.

First, they should be punctured at 5-6 different places from bottom to top to avoid water logging

Seed sowing should be carried out in the first fortnight of February. Autumn and spring sowing should be avoided as they result in poor germination and low survival.

Two seeds should be placed horizontally in the soil media. In case the seeds are placed vertically, the narrow end of the seeds should point downwards.

Conifers are shade- loving. Therefore, the container should be placed at such places so as to avoid overhead sunlight

Care during development

Irrigation should be carried out with a fountain bucket, as and when required, preferably in the morning hours before sunrise.

Waterlogging conditions should be avoided.

Weeding and cleaning should be carried out frequently

Thinning should be done just before the application of fertilisers like urea 0.5%.

Urea drenching should be carried out in the month of July-August as a top dose of fertilisers.

Seedlings should be kept in the nursery for a maximum period of 16-20 months

Diseases and pests can cause damage to the seedlings

Pre-emergence damping- off results in the death of seedling as their radicles are killed before emerging from the ground

Post- emergence damping- off appears at the collar region, forming necrotic lesions and forcing the seedlings to fall on the ground

Wilt damage occurs more under drought conditions, while root rot occurs in the succulent root tips on old seedlings

Cutworms are caterpillars that live on or below the surface of the soil. They damage new germinating seedlings at the ground level or at the collar and defoliate the leafy parts during the night.

Management

Damping off is controlled by avoiding waterlogging conditions and using raw undecomposed humus

It is also controlled by drenching the seedlings and polybags with 0.25% copper oxychloride

Wilt and root rot is controlled by drenching the soil with Carbendazim @ 0.1 % or metazyl @ 0.1 % and copper oxychloride @ 0.25 %

Control of Damage Caused by Cutworms

Flooding of polybags with water forces the larvae out of the tunnels and thereafter, they should be destroyed.

Application of soil insecticide helps in controlling damage

Forate granules, which give off an offensive odour, help in controlling insect damage

References

Dar, A. R. and Dar., G. H. 2006. Taxanomic appraisal of conifers of Kashmir Himalaya. Pakistan Journal of Biological Sciences doi 10.3923/pjbs.2006.859.867

Grassnickle, S.C. Ecophysiology of northern spruce species.The performance of planted seedlings, NRC Research Press: Ottawa, ON, Canada 2000.

Luna., R. K. 2006. Plantation forestry in India. International book distributors, 9/3, Rajput road, First floor, Dehradun, 93-100

Luna., R. K. 1996. Plantation trees. International book distributors, 9/3, Rajput road, First First floor, Dehradun, 800-900

7

Raising of Quality Planting Material of Conifers: Improved Container and Growing Media

K.N. Qaisar[1] and Monisa Banday[2]

Division of Silviculture & Agroforestry, Faculty of Forestry, Sher-e-Kashmir University of Agricultural Sciences and Technology, Benhama Ganderbal-191201, Jammu and Kashmir, India

Introduction

In India, about half of the geographical area is classified as degraded and as such, these areas are not able to produce services of any value (Khan, 1988). So, there is growing concern about the serious degradation in the quality of forests, besides the significant reduction in the tree cover in the country. To answer the prevailing conditions in the current era, one needs better nursery management to produce quality planting stock.

Forest nurseries are an essential part of forestry programmes, which cover traditional or intensively managed commercial forestry with introduced species. Field plantations should do with more productive species, but of improved quality and vigour. In order to encourage the forest tree culture, the use of quality planting stock and its subsequent management is very important. The vigour of the saplings and yield from trees depends on what we plant. One of the objectives of the National Forest Policy, 1988, is to increase the tree cover and productivity of existing forests. For the realisation of these objectives, the forest department has made large-scale efforts to bring more areas under tree cover, besides enrichment plantations in the demarcated forests. However, the success of the efforts made by the forest department and other agencies involved in the plantation programmes are not in tune with the establishment and growth of the newly planted forest species. Such situations arise due to the paucity of the quality planting material of forest species.

The importance of the quality planting material is increased manifold in the forestry sector since the land available for forestry activities is wasteland Despite having genetic superiority, most often, seedlings fail to develop into healthy saplings and trees, since their root system is not well- developed. In the last decade, forest nurseries are the most neglected part of the different forest activities. Hence, there are a large number of failures of different plantations, causing huge economic losses. The introduction of novel growing media and container technology has made a tremendous impact on forest nursery seedling production. It offers the production of a large number of healthy, uniform-sized planting stock within a short span of time under protected cultivation. It helps to improve the quality of seedlings and clonal plants since it allows growers to better manipulate mycorrhizae, biofertilisers, biopesticides and micronutrients (Will *et al.*, 2005).

Conifers, mainly *Cedrus deodara, Pinus wallichiana, Abies pindrow* and *Cupresuss torulosa,* constitute nearly 40 % of the growing stock of the state and are very important reservoirs of the fragile ecosystem and environment of the state. Nearly 50% of the stock is degraded and needs to be restocked with quality plant material of conifers.

Root Trainers

A root trainer is a specially designed cylindrical container made of opaque material with two open ends, of which, the lower end tapers gradually with a smaller open end, to provide favourable conditions to the developing root. Inside the root trainer, four to six ridges or ribs run longitudinally from one end to another, in order to prevent coiling. When a root starts to touch the ridge, it immediately changes its course and grows downward, thus avoiding coiling.

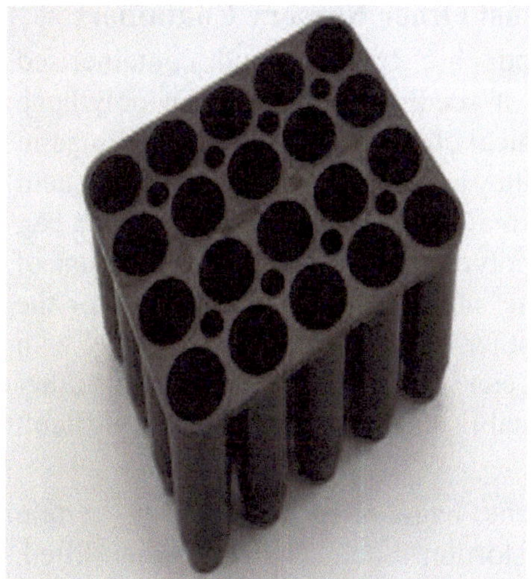

150 cc Root trainers

Japanese Cedar & Silver fir in root trainers

Deodar seedlings in root trainers

Deodar root plug

Main features of different Root Trainer capacities and sizes

Root Trainer size (cm^3)	**Top Diameter (cm)**	**Length and Width (cm)**	**Seedling growing density (per m^2)**
300 (12 cells/tray)	7	32.0 x 25.0	84
150 (25 cells/tray)	5	29.5 x 29.5	287
100 (40 cells/tray)	4	37.0 x 23.0	470

Effectiveness of Root Trainers Against Other Nursery Containers

The main seedling production methods are bareroot and containerised seedlings production. A large number of seedlings are raised in polythene bags, especially in tropical and sub-tropical climates. Raising of seedlings in polythene bags is most popular because they are easily obtained and the system is well- established (Anon, 1999 a). The major advantage of the polythene bag seedling production system is that the polythene bags carry some amount of potting mix and moisture to the planting site, which is often critical for the survival of seedlings in the field. The other advantage is that seedlings with exceptional vigour can be produced that can withstand shock better than other plants. The system lends itself to the establishment of plantations in difficult areas, such as arid and degraded sites.

However, seedlings raised in polythene bags often suffer from certain disadvantages, viz., root coiling and distortion. Such seedlings, when lifted, suffer a severe root shock, often resulting in heavy mortality. A coiled root system in a polythene bag acts as a 'time bomb' attached to the root system. The seemingly healthy plant with such a root system fails to withstand even moderate pressure, usually resulting in large-scale wind throw or similar damage. Very often, the coiled root strangulates itself as it grows, leading to stunted growth and untimely, the death of the plant (Anon, 1999 b)

Root trainers have been introduced on a reasonably large scale by various state forest departments to deal with the problems of root coiling and distortion. The most common type of root trainer is the Hiko pot with 20-25 cavities, each having 150 cc volume. The root trainer seedling production system facilitates the development of a vigorous fibrous root system (Anon, 1996). In this process, the overall surface area of the absorption zone of roots increases and when planted in the field, the seedling is expected to be established within 48 hours and starts growing quickly. It survives the prolonged drought period after planting because the development of the root system is completed in the nursery itself (Josiah and Jones, 1992).

Growing Media for Root Trainers

Anything that a plant can grow in is considered a growing medium. The growing medium consists of two or more different organic and inorganic components that are selected to lend certain physical and chemical properties to it (Chakrabarti *et al.*, 1998). A good growing medium is characterised by light weight, easy blendability, good water- holding capacity, good drainage good porosity, slight acidity, low bulk density, freedom from fungal spores etc. (Ginwal *et al.*, 2001). The proper selection of a growing medium for a particular species is important as it influences its growth quality and growth rate.

Growing medium consists of two or more different organic and inorganic components that are selected to provide certain physical and chemical properties. Mixtures of organic and inorganic components are popular because these materials have opposite, yet complimentary, physical and chemical properties. The organic component improves the physical properties, viz., porosity, CEC, water- holding capacity and maintaining a balanced ratio of carbon and nitrogen in a mix. The physical attributes may further be improved upon by the addition of inorganic components like sand, vermiculite, perlite etc. Inorganic fertilisers like urea and rock phosphate are also used at times to improve the nutrient content.

Organic components	**Inorganic components**	
Sawdust	Peat	Field soil
Manure	Vermiculite	Water
Crop by-products	Calcined clay	Sand
Coir	Polystyrene foam	Rockwool
Coconut fibre	Gravel	Composted garbage
Sphagnum moss	Air	Lava rock
Bark	Perlite	Pumice

[*Source*: Prasad, 2001]

Types of Growing Media

Root media are largely divided into those containing soil and those that do not Accordingly, the growing media have been categorised into the following two types:

1. **Soil based media:** In such media, soil is the main component. For a pot plant nursery, it is necessary to develop perfect soil-based media.

   Formulation: Traditionally, a soil-based medium is composed of equal parts by volume of loam field soil, concrete-grade sand and sphagnum peat moss or FYM amended with phosphorus and adjusted to the proper pH level. Sandy field soil is compensated by increasing the proportion of peat moss and field soil and decreasing sand, while clay soil calls for more sand.

2. **Soilless media:** In such media, soil is absent and it is formulated from various components like organic matter, peat moss, vermiculite, perlite, bark etc.

   Formulations

   i) Peat-based formulation: It ranges from 100% sphagnum or hypnum peat moss to 100% fine sand with intermediate combinations of the two, the most popular being half peat moss and half fine sand mixture

ii) **Bark-based formulation:** Bark is economically good for many areas. However, it does not possess good water and nutrient- holding capacities. Therefore, vermiculite or peat moss/perlite and/or sand are commonly added in commercial preparations of bark media. A ratio of 3 parts bark to 1 part sand to 1 part peat moss is favoured (Bilderback, 1982).

Volume/Volume Ratio	Components
1:1:1	Soil, sand, FYM
2:1:1	Soil, sand, FYM
1:1:2	Soil, sand, FYM
1:2:1	Charcoal, compost, rice husk
1:4	Charcoal, compost
2:1	Peat, Perlite
2:1:1	Peat, Perlite, Vermiculite
2:1	Peat, Sand 3:1 Peat, Sand
3:1:1	Peat, Perlite, Vermiculite
2:1:1	Peat, Bark, Sand
2:1:1	Peat, Bark, Perlite
3:1:1	Peat, Bark, Sand

Root trainers as an Eco friendly Alternative to Polythene Containers

Root trainers offer an eco-friendly approach for the mass propagation of quality plant material. The introduction of the Root trainer technology in Kashmir has had a tremendous impact on forest nursery seedling production.

The production of various forest tree species of Kashmir viz., *Cedrus deodara, Pinus wallichiana Cupressus torulosa, Abies pindrow* and *Cryptomeria japonica,* have been standardised with respect to Root trainer size and locally available growing media. The root trainer technology offers the production of a large number of healthy, uniform-sized planting stock per unit area. It also improves the quality of seedling and ease of manipulating mycorrhizae, biofertilisers and micronutrients. The design of root trainers facilitates the production of a more fibrous root system due to the open bottom, which allows air pruning of the roots. The vertical ribs in the plug prevent coiling of the roots, which is a common problem with poly bag-raised seedlings. Above all, the reusable feature of root trainers makes this technology eco-friendly. Root trainer trays can be used continuously for ten years without any breakages, unlike its polythene counterpart, which is use and throw, thus dumping a huge amount of polythene into our ecosystem each year.

Economic Viability of Root Trainers

In a study by Gera (2003), modern nursery methods of seedling production, viz., 150cc and 300cc root trainers, improved polythene bag seedling production, where the bags were placed on mounted angle iron beds. The results were compared with those of the conventional nursery production system by raising seedlings of four tree species, viz., *Acacia catechu, Albizia lebbek, Azadirachta indica* and *Pinus roxburghii*. A comparative study was made on the basis of the costs involved and benefits received. In order to bring commonality of comparisons, the study involved raising 8000 seedlings every year using selected seedling production systems on a continuous basis for ten years. The results have revealed that the seedlings raised on mounted angle iron (MAI) beds recorded not only appreciable values of height and collar diameter, but also gave better values on seedling quality parameters. Also, the seedlings raised on MAI beds were found to be most cost-effective when compared with other nursery systems on the basis of benefit- cost analysis.

Benefit-cost ratios of the different seedling production systems

S.No.	Seedlings Production System	Benefit cost ratio	
		12% discount rate	6% discount rate
1.	Conventional nursery system	0.68	0.68
2.	Seedlings kept on MAI beds	0.99	1.03
3.	150 cc root trainers	0.79	0.84
4.	300 cc root trainers	0.67	0.73

Benefit-cost ratio of different root trainer systems (Bashir et.al., 2009)

Root Trainer size (cm^3)	Benefit-Cost ratio (at 6% discount rate)
300	1.41
150	1.54
100	1.61

Raising of Conifer Seedlings in Root Trainers/Perforated Polybag

Deodar: For raising quality nursery stock, seedlings should be raised in bottom- perforated polythene bags (9" x 4") filled with Soil + Sand + FYM + Dalweed in the ratio of 1:2:3:2

Kail: For raising quality nursery stock, seedlings should be raised in 300 cc root trainers filled with Soil + Sand + Forest litter in the ratio of 1:1:2

Silver Fir: FFor raising quality nursery stock, seedlings should be raised in 300 cc root trainers filled with peat moss + vermiculite in the ratio of 1:1

Himalayan: Cypress For raising quality nursery stock, seedlings should be raised in 250 cc trainers filled with Soil + Sand + FYM + Dalweed in the ratio of 1:2:3:2

Japanese Cedar: For raising quality nursery stock, seedlings should be raised in 300 cc root trainers filled with peat moss+ vermiculite in the ratio of 1:1

Field Performance of Certain Conifers

1. ***Cedrus Deodara***

   Deodar is the prime timber species of Kashmir, where the famous houseboats are built from it. Deodar seedlings raised in bottom perforated poly bags (9"x4") with three pairs of bottom perforation performed best under field conditions in terms of survival % (80), height (77.15 cm), collar diameter (15 mm) and increments of 33.75 cm in height and 5.49 mm in collar diameter during a period of two years.

   The field performance of root trainer (150cc) seedlings was observed to be better as compared to normal poly bag seedlings. Poly bag seedlings recorded a survival rate of only 40%, height (46.74 cm), collar diameter (8.35 mm) and increments of 7.38 cm in height and 2.09 mm in collar diameter during a period of 2 years.

   Growing medium M_2 (Soil: Sand: FYM: Dalweed:: 1:2:3:2) resulted in the best field performance of Deodar in terms of survival (64.25%), height (62.3 cm) and collar diameter (13.5 mm) (Qaisar, et.al., 2008 a)

2. ***Cupressus Torulosa***

   The nursery performance of Himalayan Cypress (*Cupressus torulosa*) and Deodar (*Cedrus deodara*) was evaluated with four containers viz., C_1 (Normal polybags), C_2(Bottom-perforated polybags), C_3(Root trainer 150 cc), C (Root trainer 250cc) and three growing media as M_1(Soil, Sand, FYM :: 1:2:3), M_2 (Soil, Sand, FYM, Dalweed :: 1:2:3:2) and M_3(Soil, Sand, FYM and Rice husk :1:2:3:2).

   The best container for raising quality nursery stock of Cypress & Deodar was normal and bottom-perforated polybags (9"x4") respectively. Root trainers of 250cc can be used for obtaining better quality parameters, viz., sturdiness (low), root/shoot ratio and ratio of fibrous/lateral root for Deodar and *Cupressus* seedlings. The best growing medium for raising quality nursery stock of Deodar and *Cypressus* was Soil: Sand: FYM: Dalweed (1:2:3:2) (Qaisar *et al.*, 2008 b

3. ***Pinus Wallichiana***

   The field performance of Blue pine seedlings were raised in two sizes of root trainers and four growing media and observed for two years.

The seedlings raised in Root trainer size 300 cc (C) recorded maximum field survival 85.00%, height (18.75 cm) and collar diameter (4.88 mm). Seedlings raised in the growing medium M (Soil:Sand:Forest Litter) recorded maximum field survival (100.00 %), height (20.00 cm) and collar diameter (5.22 mm). The interaction of C_2 M_4 (Root Trainer 300 cc + Soil: Sand: Forest Litter) recorded maximum survival (100.00%), height (20.66 cm) and collar diameter (5.51 mm).

4. ***Abies Pindrow***

An experiment was conducted of C_2M_4 (*Abies pindrow*) seedlings raised in five different growing medium in 300cc root trainers-M_1 (Peat moss + vermiculite, 1 : 1), M_2 (Dal weed +Sand, 1:1), M_3 (*Cupressus* leaf litter+ Sand, 1:1), M_4 (Peat moss+Sand, 1:1) and M_5 (Soil + Sand + FYM, 1:1:1). The growth of seedlings was recorded after two years and six months of sowing. The growing medium M_1 (Peat moss + Vermiculite, 1:1) recorded maximum height (10.2 cm), collar diameter (3.71 mm) and fresh shoot and root weight (3.79 and 4.22g) respectively, similarly the seedlings raised in M_1 also recorded maximum height increment of 3.92 cm (Qaisar *et al.*, 2009).

References

Anon, (1996). Field guide on improved nursery technology-Vision 2000, A.P. Forest Department.

Anon, (1999 a). Improved polythene bag seedling production system. J&K SFRI publication No. SFRI/SDD (Jm)-4.

Anon, (1999 b). Root trainer seedling production system for enhancement of forestry plantations. J&K SFRI publication No. SFRI/SDD (Jm)-3.

Bashir,A.,Qaisar,K.N.,Khan,M.A. and Majeed, M.2009.Benefit- cost analysis of raising Pinus wallichiana seedlings in different capacities/sizes of root trainers in the nursery.For. Stud.China,vol.11(2):118-121

Bilderback, T.E. 1982. Container soils and soil less media. In: Nursery Crop Production. Raleigh, NC; North Carolina State University, Agricultural Extension Service. Chakrabarti, K., Zaidi, A. and Barari, S. 1998. Compost for container nursery – A West Bengal experience. Indian Forester 124(1):17-30.

Gera, M., Singh, A.K. and Chand,P.2003.Cost effectiveness of different containerized nursery technologies. Indian Forester, 129 (10):1201-1210

Ginwal, H.S., Rawat, P.S., Bhandari, A.S., Krishnan, C. and Shukla, P.K. 2001. Selection of proper potting mixture for raising Acacia nilotica seedlings under root trainer seedling production system. Indian Forester 127(1) : 1239-1250.

Josiah, S.J. and Norman Jones (1992). Root trainers for seedlings production system for tropical forestry and agroforestry. Land resource series No. 4, World Bank.

Khan,P.A.,Qaisar,K.N.,Khan,M.A.and Mugloo,J.A.2008.Nursery growth and survival of containerized Japenese cedar (Cryptomeria japonica) seedlings raised in different growing media. SKUAST J.Res.10:139-142.

Prasad, S. and Kumar, U. 2001. Root media In: Greenhouse Management for Horticultural Crops. Agrobios (India) 112-143.

Qaisar, K.N., Khan, P.A., Khan, M.A., J.A.Mugloo and Rather, T.A.2009.Growth performance of root trainer grown Silver Fir (Abies pindrow) seedlings in different growing media. Indian Forester.135(4):554-558

Qaisar,K.N., Khan,P.A. and Khan, M.A.2008.Container type and growing media for raising quality seedling stock of Cedrus deodara. Indian Jr. Forestry, vol.31 (3): 383-388.

Qaisar,K.N.,Khan,P.A. and Khan,M.A.2008.Nursery performance of Cupressus torulosa seedlings raised in different containers and growing media. Environment & Ecology 26(4) :1505-1508

Will, E. and James, E. 2005. Growing media for green house production. Agricultural Extension Series PB 1618.

8

Application of Microbial Inoculants for Raising Quality Forest Nursery Stock / for Quality Forest Nursery Raising

Malik Asif, A.H. Mughal, Zaffar Mehdi, Bisma R., Saima S., Misbah A. and M.A. Malik

Division of Basic Sciences & Humanities, FOA, Wadura, SKUAST-Kashmir Jammu and Kashmir, India

Introduction

Biofertilisers are defined as preparations containing the living cells or latent cells of efficient strains of microorganisms that assist in the uptake of nutrients in crop plants by their interactions in the rhizosphere, when applied through seed or soil. They accelerate certain microbial processes in the soil, which augment the extent of the availability of nutrients in a form easily assimilated by plants. The increasing pressure of human and livestock population, indiscriminate extraction of forest produce, regular forest fires and mining activities have resulted in soil erosion, loss of fertility and moisture content and decreasing productivity of forests, which pose manifold problems for the restoration of the ecosystem. Thus, the microbial inoculants present in the soil form a strong and important component of our soils, mainly owing to their role in promoting plant growth by providing access to the nutrients, nitrogen fixation, mobilisation of some unavailable nutrients and production of antifungal antibiotics. The indiscriminate use of inorganic fertilisers and pesticides is neither environmentally safe nor economically feasible. There is a pressing demand for microbial inoculants for quality seedling production in nurseries and also the establishment of plantations to increase the forest productivity. Bioinoculants are cost-effective, eco-friendly, cheaper and renewable sources of plant nutrients and play a vital role in maintaining long-term soil fertility and sustainability. Moreover, they form an important component of organic farming practices. Thus, to meet challenges like poor regeneration, deforestation and spread of wastelands, the introduction of microbial inoculants at the nursery stage of forest trees has become crucial. Various aspects of the mycorrhizal impact of forest trees have been studied. In tropical countries like India, there is a problem of nitrogen deficiency and phosphorous non-availability, especially in degraded or stressed soils. In order to overcome these deficiencies, chemical fertilisers are being used. However, the indiscriminate use of chemical fertilisers, regardless of climatic, soil and other factors, has affected the environmental quality and soil ecosystem. There are many reports that highlight the adverse effects of inorganic fertilisers on both beneficial micro flora and micro fauna as well as their activities (Zargar *et al.*, 1999).

Like other soils, forest soils are also facing an acute shortage of essential nutrients mainly due to continuous erosion. Moreover, the fertilisation of forest soils is not easy as forests occupy remote, less accessible and less fertile lands. It is, therefore, feasible to make use of native microbial inoculants as natural nutrient mobilisers, instead of inorganic fertilisers, for better growth, development and survival of tree species. Simultaneously, the use of microbial inoculants will reduce the use of synthetic fertilisers, which in turn, will decrease the cost of cultivation and ill- effects on human health and soil ecosystem, besides improving the growth of plants significantly. Thus, the establishment of suitable microbial inoculants is essential to improve the survival and quality of planting stock so as to undertake national developmental programmes of afforestation, reforestation, wasteland-reclamation and social forestry successfully. The present study reviews the impact of various microbial inoculants on plant growth as well as the chemical and microbiological characteristics of forest soils.

Effect of Microbial Inoculants on Plant Growth Characteristics

Microorganisms constitute an integral and significant constituent of soil. There is a vast array of microorganisms in the soil which play their role in improving the soil health as well as plant growth. A few microorganisms that play a highly significant role are discussed hereunder:

Effect of *Azotobacter* and *Azospirillum* spp.

Azotobacter was first isolated and described by Beijernick. Other species related to *Azotobacter* are known as *Beiferinckiaindica* and *Derxiagummosa*. In general, *Azotobacter* is gram-negative, polymorphid, rod-shaped and varying from 2.0 to 7.0 x 1.0 to 2.5 μ. Occasionally, an adult cell may measure up to 10-12 μ. However, the morphology of cells is dependent upon the composition of the medium, species, strain, age of culture and growth conditions. The young cells have *Peritrichous flagella* which serve as locomotory organs. The beneficial effects of *Azotobacter* and *Azospirillum* in terms of improved growth characteristics of various crops is attributed to their nitrogen fixing ability, synthesis of growth promoting substances like indoleacetic acid, gibberellic acid, Vitamin B and antibiotics, besides enhancement in the branching of roots and uptake of No_3 -, NH_4+, H_2PO_4 -, K+, Rb++ and Fe++ (Ahlawat *et al.*, 1995). Moreover, these species have also been reported to inhibit the hatching of egg masses of nematodes, thereby improving the plant growth (Chahal and Chahal, 1988). Microbial inoculants colonise the rhizosphere and stimulate plant growth by providing necessary nutrients. Thus, they enhance plant growth by providing plant growth promoting substances and increasing the nutrient availability in the root zone (Jackobsen *et al.*, 1994).

Effect of *Pseudomonas fluorescensc and Bacillus subtillis*

Among the various antagonists of fungal pathogens, *Pseudomonas* and *Bacillus* spp. are the most important plant growth promoting rhizobacteria. They offer an environmentally sustainable approach to increase crop production and health. They have been found to cause significant improvement in the plant growth of many plants. PGPR are also associated with the production of antibiotics, hormones, siderophores and HCN, as well as with molecular mechanisms, which can stimulate induced systemic resistance in plants. It is interesting to note that these mechanisms do not affect the mycorrhizal fungi with which PGPR are associated (Edwards *et al.*, 1999). Dominghez *et al.,* (2011) while studying the seedlings of *Pinus halpensis* inoculated with mycorrhizal fungus *Tuber melanosporum* and the rhizo bacterium, *Pseudomonas fluorescens* CECT 844 under non-limiting greenhouse conditions, reported that the inoculations improved the growth and nutrient uptake of the seedlings. Although the combination of *Pseudomonas fluorescens* CECT 844 and *T. melanosporum* did not generally lead to a significant improvement over the positive effects of a simple inoculation of *T. melanosporium*, the addition of *P. fluorescens* CECT 844 doubled the rate of the mycorrhization of *T. melanosporium*.

Effect of *Pisolithus Tinctorius* and *Laccaria Laccata*

Mycorrhizal fungi have received considerable interest from researchers, particularly for the past 40 years or so, when their role as modulators of plant growth began to be universally recognised. It is only more recently that their interactions with other soil organisms have been studied in more detail. In the natural ecosystem, hundreds of ectomycorrhizal fungi have been reported on a single tree species and approximately, 70% plants are obligatory dependent on fungal associates, 12% facultatively dependent and 18% remain typically uninfected (Trappe, 1987). It has been estimated that more than 2100 species of fungi form an ectomycorrhizae association with forest trees in North

America alone, while there are over 5000 species of fungi that can form ectomychorrhizae on about 2000 species of woody plants worldwide (Marx, 1969). Pavan (2011), while working on the influence of bioinoculants like AM fungi, Rhizobium, Azotobacter and PSB on the growth of nursery seedlings of *Albizia lebeck,* reported that mycorrhizal association and resting spore population showed an increasing trend till the age of 180 days and the dual inoculations were found to be more effective for both AM colonisation and biomass production. It was further reported that mixed inocula might be used more efficiently than single inoculum for field inoculation. Anabela (2008) reported that at 50 days of mycorrhizal induction in *Castanea sativa* plants, growth was higher for mycorrhizal plants than for non-mycorrhizal ones. It was further observed that the length of the major roots was lower in mycorrhizal plants after 40 days, while fresh and dry weights were higher in mycorrhizal plants after 30 days.

Types of Biofertilizers

1. Rhizobium
2. Azotobacter
3. Azospirillum
4. Cyanobacteria
5. Azolla
6. Phosphate solubilising microorganisms (PSM)
7. AM fungi
8. Silicate solubilizing bacteria (SSB)
9. Plant Growth Promoting Rhizobacteria (PGPR)

Rhizobium: Rhizobium is a soil habitat bacterium, which is able to colonise the legume roots and fixes the atmospheric nitrogen symbiotically. The morphology and physiology of rhizobium will vary from free-living conditions to the bacteroid of nodules. They are the most efficient biofertilisers as far as the quantity of nitrogen fixed is concerned. They have seven genera and are highly specific to form nodules in legumes, referred to as cross- inoculation groups. Rhizobium inoculant was first made in USA and commercialised by private enterprises in the 1930s.

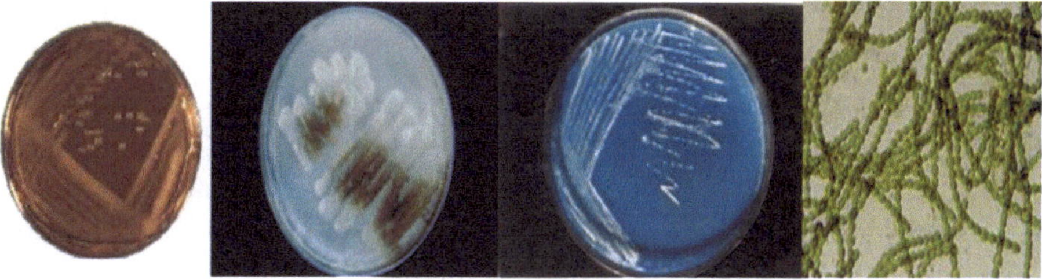

Fig. 1. Cultures of different inoculants

Azotobacter: Of the several species of Azotobacter, *A. chroococcum* is the dominant inhabitant in arable soils, capable of fixing N2 (2-15 mg N2 fixed /g of carbon source) in culture media.

The bacterium produces abundant slime which helps in soil aggregation. The numbers of *A. chroococcum* in Indian soils rarely exceeds 105/g soil due to lack of organic matter and the presence of antagonistic microorganisms in the soil.

Azospirillum: *Azospirillum lipoferum* and *A. brasilense* are primary inhabitants of the soil, rhizosphere and intercellular spaces of the root cortex of graminaceous plants. They have an associative symbiotic relation with the graminaceous plants. The bacteria of Genus Azospirillum are N2 fixing organisms, isolated from the root and above ground parts of a variety of crop plants. They are Gram- negative. Apart from nitrogen fixation, growth promoting substance production (IAA), disease resistance and drought tolerance are some of the additional benefits of Azospirillum inoculation.

Phosphate solubilising microorganisms (PSM): Several soil bacteria and fungi, notably species of *Pseudomonas, Bacillus, Penicillium, Aspergillus* etc., secrete organic acids and lower the pH in their vicinity to bring about the dissolution of bound phosphates in the soil. Increased yields of wheat and potato were demonstrated due to the inoculation of peat- based cultures of *Bacillus polymyxa* and *Pseudomonas striata*. Currently, phosphate solubilisers are manufactured by agricultural universities and some private enterprises and sold to farmers through governmental agencies. There appear to be no checks on either the quality of the inoculants marketed in India or the establishment of the desired organisms in the rhizosphere.

AM fungi: The transfer of nutrients, mainly phosphorus and also zinc and sulphur, from the soil to the cells of the root cortex is mediated by intracellular obligate fungal endosymbionts of the genera Glomus, Gigaspora, Acaulospora, Sclerocysts and Endogone, which possess vesicles for the storage of nutrients and arbuscles for chanelling these nutrients into the root system. By far, the most common genus appears to be Glomus, several species of which are distributed in the soil.

Silicate solubilizing bacteria: Microorganisms are capable of degrading silicates and aluminium silicates. During the metabolism of microbes, several organic acids are produced and these have a dual role in silicate weathering. They supply H+ ions to the medium and promote hydrolysis and organic acids like citric acid, oxalic acid, keto acids and hydroxycarbolic acids, which form complexes with cations, promote their removal and retention in the medium in a dissolved state.

Plant Growth Promoting Rhizobacteria: The group of bacteria that colonise roots or rhizosphere soil and are beneficial to crops are referred to as plant growth promoting rhizobacteria (PGPR).

The PGPR inoculants that are currently commercialised seem to promote growth through at least one mechanism - suppression of plant disease (termed Bioprotectants), improved nutrient acquisition (termed Biofertilisers) or phytohormone production (termed Biostimulants). Species of *Pseudomonas* and *Bacillus* can produce, (as yet, not?) well-characterised phytohormones or growth regulators that cause horticulture and forest crops to have greater amounts of fine roots, which have the effect of increasing the absorptive surface of plant roots for the uptake of water and nutrients. These PGPR are referred to as biostimulants and the phytohormones they produce include indoleacetic acid, cytokinins, gibberellins and inhibitors of ethylene production.

Benefits of Liquid Biofertiliser Technology

The advantages of Liquid Biofertilisers over conventional carrier based Biofertilizers are listed below:

- Longer shelf life, viz.,12-24 months.
- No contamination and no loss of properties due to storage up to 45° C.
- Greater potential to fight against native population.
- High populations of more than 109 cells/ml can be maintained from 12 to 24 months.
- Easy identification by typical fermented smell.
- Cost saving on carrier material, pulverisation, neutralisation, sterilisation, packing and transport.
- Quality control protocols are easy and quick and better survival on seeds and soil.

- No need of running biofertiliser production units throughout the year.
- Very easy to use by the nursery growers and farmers.
- Dosage is 10 times less than carrier- based powdered biofertilisers.
- High commercial revenues and high export potential.
- Very high enzymatic activity since contamination is nil

Methods of Applications

There are three ways of using Liquid Biofertilizers

1. Seed treatment
2. Root dipping
3. Soil application

Seed treatment: Seed treatment is the most common method adopted for all types of inoculants. It is effective and economic. For small quantity of seeds (up to 5 kg), the coating can be done in plastic bags.

Root dipping: This method is used for the application of most inoculants in forest nurseries. The required quantity of inoculants has to be mixed with 5-10 litres of water at one corner of the nursery and the roots of the seedlings have to be dipped in it for a minimum duration of 30 mins. before transplantation.

Soil application: Use 200 ml of PSM per acre. Mix PSM with 400 to 600 kg of cow dung FYM, along with half a bag of rock phosphate, if available. The mixture of PSM, cow dung and rock phosphate has to be kept under any tree or in the shade overnight and 50% moisture has to be maintained. Use the mixture as soil application in rows or during the levelling of the soil.

Dosage and application of liquid Bio-fertilizers in different crops

Crop	Recommended Biofertiliser	Application method	Quantity to be used
Deodar, Cypress & Blue pine seedlings	N-Fixers, P-Solubilisers, KSB,VAM	Seedling/soil 1 treatment	20 ml/seedling 1 ltr/acre of nursery
All fruit/agro-forestry (herb, shrubs, annuals and perennials) plants for fuel wood fodder, fruits,gum,spices,leaves,flowers, nuts and seeds purpose	Azotobacter	Seedling/ Soil treatment	20 ml/plant at nursery 1 ltr / acre of nursery

Conclusion

The application of microbial inoculants viz. *Pisolithustinctorius, Laccarialaccata, Azotobacter, Azospirillum, Pseudomonas fluorescencs* and *Bacillus subtilis* is most effective and efficient at nursery stage of most temperate, sub-tropical and tropical forest species.Their application improves plant growth characteristics, seedling survival, chemical and microbiological characteristics of the soil. Environmental stresses are becoming a major problem and productivity is declining at an unprecedented rate. Our dependence on chemical fertilisers and pesticides has encouraged the thriving of industries that are producing life-threatening chemicals and which are not only hazardous for human consumption but can also disturb the ecological balance.The new technology developed viz; biofertiliser technology has a wide scope for application in the forest ecosystems.

References

Ahlawat, V.P., Suneel, S. and Kartar, S. 1995. Plant Microbe Interaction in Sustainable Agriculture. [Eds R.K. Behl, A.L. Khurana and R.C. Dogra]. CCS HAU, Hisar and MMB, New Delhi, pp. 121-131.

Anabela, M. 2008. In vitromycorrhization of micropropagatedplants : studies on Castaneasativa Mill. Sustainable Agriculture and Forestry10 : 319-334.

Chahal, P.P.K. and Chahal, V.P.S. 1988. Biological control of root-knot nematode of brinjal with Azotobacterchroococcum. In :Advances in Plant Nematology [Eds. M.A. Maqbool, A.M. Golden, A.M. Gaffar and A. Krusberg]. Research Centre, University of Karachi, pp. 257-263.

Dominguez, J.A., Martin, A., Anriguez, A. and Albanesi, A. 2011. The combined effects of Pseudomonas fluorescensand Tuber melanosporum on the quality of Pinushalepensis seedlings. Mycorrhiza(10(5) : 1007-1016.

Edwards, S.G., Young, J.P.W. and Fitter, A.H. 1999. Interactions between Pseudomonas fluorescensbiocontrol agents and Glomusmosseae, an arbuscularmycorrhizal fungus, within the rhizosphere. Microbiology Letters166 : 297-303.

Jackobsen, I., Joiner, E.J. and Larsen, J. 1994. Hyphal phosphorus transport, a keystone to mycorrhizal enhancement of plant growth. In :Impact of AM on Sustainable Agriculture and Natural Ecosystem. [Eds. S. Gianinazzi and H. Schuepp]. BirkhauserVerlag, Switzerland, pp. 133-146.

Marx, D.H. 1969. The influence of ectotrophicmycorrhizal fungi to root pathogenic fungi and soil bacteria. Phytopathology59 : 153-163.

Pavan, K.P. 2011. Influence of bioinoculants on the growth of Albizialebeck in nursery conditions. Research Journal of Agricultural Sciences2(2) : 265-268.

Trappe, J.M. 1987. Phylogenetic and ecological aspects of mycotrophy in the angiosperms from an evolutionary standpoint. In :Ecophysiology of VA mycorrhizal Plants. [Ed. S.R. Safir]. CRC Press, Boca Raton, USA, pp. 2-25.

Zargar, M.Y., Dar, G.H., Zargar, A.H. and Ganai, B.A. 1999. Effect of urea on rhizospheremicroflora of common bean (Phaseolus vulgaris). SKUAST Journal of Research 1 : 52-55

9

Vermicompost Technology and Its Application in Forest Nursery Raising

Malik Asif, A.H. Mughal, Zaffar Mehdi, Saima S., M.A. Malik and Bisma R.

Division of Basic Sciences & Humanities, FOA, Wadura, SKUAST-Kashmir

Vermicompost is a nutrient-rich, microbiologically-active organic amendment that results from the interaction between earthworms and microorganisms during the breakdown of organic matter. It is a stabilised, finely divided peat-like material with a low C:N ratio, high porosity and high water-holding capacity, in which, most nutrients are present in forms that are readily taken up by plants. Earthworms act as mechanical blenders, and by fragmenting the organic matter, they modify its physical and chemical status by gradually reducing the ratio of C:N and increasing the surface area exposed to microorganisms, thus making it much more favourable for microbial activity and further decomposition. The use of compost in forestry and horticulture has occasionally been shown to be limited by the high electrical conductivity and excessively high amount of certain ions that cause phytotoxicity, as a consequence of the chemical properties of the initial waste and /or inadequate composting procedures. These adverse effects, although possible, are less likely to occur when vermicompost is used as a potting amendment, especially for raising forest nurseries. The production of high quality forest tree seedlings in nurseries is very important as far as tree nurseries and farmers are concerned, for reversing the current degradation of natural forests, woodlands and scrublands. Quality seedling production is the main objective of forest nurseries, but the slow growth of seedlings limits the high quality seedling production. The readiness of seeds to germinate for further multiplication is much warranted and there is a need to optimise a growing media for high quality seedling production in tree nurseries. The slow-growing nature of seedlings is the major limiting factor for successful seedling production in tree nurseries, which can be enhanced by standardising the appropriate growing medium with the addition of quality vermicompost.

The production of high quality forest tree seedlings in nurseries is very important as far as tree nurseries and farmers are concerned, for reversing the current degradation of natural forests, woodlands and scrublands. Quality seedling production is the main objective of forest nurseries, but the slow growth of seedlings limits the high quality seedling production. The slow-growing nature of seedlings is the major limiting factor for successful seedling production in tree nurseries, which can be enhanced by standardising the appropriate growing medium with the addition of quality vermicompost. This can improve the physical properties of the medium like aeration, drainage and water-holding capacity, which are otherwise lacking. The nutrient value of the growing media can be enriched by the addition of specific nutrients and pure cultures of beneficial microbes, like plant growth promoting microorganisms (PGPR), capable of enhancing the availability of nutrients for plant growth. Animal manure is a valuable resource as a soil fertiliser because it provides large amounts of macro and micronutrients for crop growth and is a low-cost, environment- friendly alternative to mineral fertilisers. The processing of this waste material through controlled biooxidation processes, such as composting, reduces the environmental risk by transforming the material into a safer and more stable product which is suitable for application to the soil (Lazcano *et al.*, 2008). As compared to mineral fertilisers, compost produces a significantly greater increase in soil organic carbon and some plant nutrients (Garcia-Gil *et al.*, 2000, Bulluck *et al.,* 2002). Long-term beneficial effects of composted materials are also observed in soil humic substances, due to an increase in the complexity of their molecular structure which increases the humic/ fulvic acid ratio, as well as in soil absorption properties (with increased cation exchange capacity and base saturation) (Weber *et al.*, 2007). The vermicompost expression is a term which is used for the final product (humus-like material) of the composting procedure of organic waste materials by soil worms. Many organic wastes are converted into worm manure (vermicompost) by different species of earthworms. These include cow manure, horse waste, olive leaves, paper waste, sheep/goat manure, ground rice waste and tea/coffee waste. The increase in the human population, industrialisation and agricultural practices have led to an increased accumulation of wastes. Cow manure and vegetal wastes are most widely used as bed ingredients in the closed system for the production of vermicompost. Vermicompost fertilisers are an important component of sustainable nursery forest raising and are used in many countries all over the world.

Vermicomposting Technology

Vermicomposting technology is a novel and eco-friendly technology with no adverse effect on the immediate ecosystems. Vermicompost is a nutrient-rich, microbiologically-active organic amendment that results from the interaction between earthworms and microorganisms during the breakdown of organic matter. It is a stabilised, finely divided peat-like material with a low C:N ratio, high porosity and high water-holding capacity, in which, most nutrients are present in forms that are readily taken up by plants (Dominguez, 2004). Earthworms are the crucial drivers of the process as they accelerate and fragment the substrate, thus drastically altering the microbial activity. Earthworms act as mechanical blenders and by fragmenting the organic matter, they modify its physical and chemical status by gradually reducing the ratio of C:N and increasing the surface area exposed to microorganisms, thus making it much more favourable for microbial activity and further decomposition (Dominguez *et al.*, 2010). As a result of the different processes involved in the production of compost and vermicompost, they exhibit different physical and chemical characteristics that affect soil properties and plant growth in diverse ways. Vermicomposting generally converts organic matter to a more uniform size, which gives the final substrate a characteristic earthy appearance, whereas the material resulting from composting usually has a more heterogeneous appearance (Tognetti *et al.*, 2005). The adverse effects, although possible, are less likely to occur when vermicompost is used as a potting amendment (Chaoui *et al.*, 2003).

Earthworms

Earthworms are capable of transforming garbage into'gold'. Charles Darwin described earthworms as the 'unheralded soldiers of mankind' and Aristotle called them the 'intestine of the earth', as they could digest a wide variety of organic materials (Darwin and Seward, 1903). Soil volume, microflora and fauna are influenced by earthworms and have been termed as 'drilosphere'. The soil volume includes the external structures produced by earthworms, such as surface and below ground casts, burrows, diapause chambers as well as the earthworm body surface and structures associated with the internal gut in contact with the soil (Lavelle *et al.*, 1989; Brown *et al.*, 2000). The intestine of earthworms contains a wide range of microorganisms, enzymes and hormones, which aid in the rapid decomposition of half-digested material, transforming them into vermicompost in a short time (nearly 4–8 weeks) (Ghosh *et al.*, 1999) as compared to the traditional composting process, which uses only microbes and therefore, requires a prolonged period (nearly 20 weeks) for compost production (Bernal *et al.*, 1998). As the organic matter passes through the gizzard of the earthworm, it is ground into a fine powder, after which, the digestive enzymes, microorganisms and other fermenting substances act on it, further aiding its breakdown within the gut. It is finally passed out in the form of 'casts', which are later acted upon by microbes in the gut of earthworms, converting them into the mature product, the 'vermin composts'.

Earthworm Species Suitable for Vermicompost Production

The following earthworm species may be used for preparation of vermicompost:

1. *Eisenia foetida (red worm)*
2. Eudrilus eugeniae (night crawler)
3. *Perionyx excavates*

Production Technology of Vermicompost

Raw materials required: Cow dung and any other biodegradable waste, such as crop residue, forest leaf litter, weed biomass, vegetable waste, slaughter house waste, biodegradable portion of urban and rural waste etc. may be used for the preparation of vermicompost.

Methods of preparation: The Vermicompost can be prepared in locally made pits.

- The standard size of the pits should be 8 x 3 x 3ft (l x b x d).
- The available bio-wastes are to be collected and heaped under the sun for about 3 weeks and then, covered with a shade net for the pre-decomposition process.
- The heap can be sprinkled with cow dung slurry.
- A thin layer of half- decomposed cow dung (12 inches) is to be placed at the bottom.
- Place the chopped weed biomass/ partially decomposed leaf litter and cow dung layer- wise (10-20 cm) in the pits up to a depth of 3 ft.
- The bio- waste and cow dung ratio should be 60: 40 on dry weight basis.
- Release about 2-3 kg earthworms per pit.
- Place wire net / shade net over the pits to protect the earthworms from birds.
- Water should be sprinkled to maintain 70-80% moisture content.
- Turning of the bio-waste material should be done at an interval of 15 days.
- The provision of a shade over the compost is essential to prevent the entry of rainwater, snow and direct sunshine.
- The sprinkling of water should be stopped when 90 % bio-wastes are decomposed. The maturity could be judged visually by observing the formation of the granular structure of the compost at the surface of the pit.

- Harvest the vermicompost by scraping layer- wise from the top of the pit and heap it under shade. This will help in the separation of earthworms from the compost. Sieving may also be done to separate the earthworms and cocoons.

Separation of Earthworms and Cocoon

- Heap the harvested vermicompost for 6-12 hours under shade for separation of the earthworm.
- Sieve the vermicompost for the separation of baby earthworm and cocoons.
- Dry the vermicompost (if necessary) under shade to keep the moisture content below 20%.
- Separate the earthworms and cocoons for reuse.

Protection from enemies

- Bio-wastes that are free from ants /termites etc. are to be used for vermicompost preparation.
- The vermicompost thus prepared would normally have the nutrients of the following concentrations (Table 1).

Advantages of vermicompost

- Vermicompost is a rich source of vitamins, hormones, enzymes, macro and micronutrients, which when applied to plants, help in efficient growth (Prabakaran, 2005).
- The vermiculture provides for the use of earthworms as natural bioreactors for cost- effective and eco- friendly waste management (Aalok *et al.*, 2008).
- Vermicompost has an excellent effect on overall plant growth, encourages new shoots/ leaves and improves the quality and shelf life of the produce.
- Vermicompost is free flowing, easy to handle, store and apply and does not have bad odour.
- Earthworms change the soil in many beneficial ways. They increase the soil's nutrient content available for plants (nitrates, phosphates, exchangeable calcium and soluble potassium), growth regulators and useful bacterial populations (Bhadauria and Rama Krishnan, 1996).
- Vermicompost improves soil structure, texture, aeration, water holding capacity and prevents soil erosion.
- Vermicompost is rich in beneficial microflora, such as N-fixers, P-solubilisers, cellulose decomposing micro-flora etc.

- The digestive enzymes of earthworms are responsible for the decomposition and humification of organic matter. These enzymes are active at a very narrow pH range and efficiently maintain the non – linear pH parameters (Gajalakshmi and Abbasi, 2003) Vermicompost is rich in several enzymes and growth regulators such as auxins, gibberellins etc.
- Vermicompost contains earthworm cocoons and increases the population and activity of earthworms in the soil.
- Vermicompost prevents the loss of nutrients and increases the use efficiency of chemical fertilisers.
- Vermicompost is free from pathogens, toxic elements, weed seeds etc. Vermicompost minimises the incidences of pests and diseases in crops.
- The availability of nutrients from vermicompost is faster due to its narrow C: N ratio.
- Vermicompost enhances the decomposition of organic matter in the soil

Table 1. Nutritional Composition of Vermicompost

Nutrients	**Percentage (%)**
Nitrogen	1.9
Phosphorus	1.18
Potassium	1.88
Calcium	0.5 -1.0
Magnesium	0.2 -0.3
Sulphur	0.4 – 0.5
Iron	0.8 -1.5
Copper (ppm)	18
Zinc (ppm)	100

Dosage: The dosage of vermicompost depends upon the type of crop grown in the field/nursery. For fruit crops, it is applied in the tree basin. It is added in the pot mixture for potted ornamental plants and for raising forest nurseries.

Plant	**Dose/rate**
Pots/Polybags	100-200 g/pot
Fruit crops	3-5 kg/plant
Field crops	5-6/ha

Vermicomposting Derived Liquids

If compared with the conventional composting process, the beauty of vermicomposting is that the time span for stabilising and processing the waste is short, as it does not undergo the thermophilic phase of composting. In the vermicomposting process, earthworms will ingest the substrate introduced into the reactor. There are few terms that can be found in describing vermicomposting derived liquids. The common ones are:

1. Vermiwash
2. Vermicomposting leachate
3. Vermicompost aqueous extracts

Effects of Vermicompost on Plant Growth

Vermicompost significantly stimulates the growth of a wide range of plant species including several ornamental hedges and horticultural crops. The positive effects of vermicompost have also been observed in forestry species such as acacia, eucalyptus and pine tree (Lazcano *et al.*, 2010a, 2010b). Vermicompost has been found to have beneficial effects when used as a total or partial substitute for mineral fertiliser in peat-based artificial greenhouse potting media and as soil amendments in field studies. Variations were observed in an experiment studying the effects of vermicompost and vermicompost extracts on the germination and early growth of six different progenies of maritime pine (Lazcano *et al.*, 2010a). In this experiment, the speed of maturation increased, relative to the control without vermicompost, in three out of the six pine progenies, decreased in two and was unaffected in the last progeny. It may be expected that different hybrids or plant genotypes will respond differently to vermicompost, considering that plant genotype determines important differences in nutrient uptake capacity, nutrient use efficiency and resource allocation within the plant. Different genotypes may, therefore, enhance root growth or modify root exudation patterns in order to increase nutrient uptake (Cavani and Mimmo, 2007) and all these strategies will determine the establishment of different interactions with the microbial communities at the rhizosphere level. Vermicompost has also been found to have a wide range of indirect effects on plant growth, such as the mitigation or suppression of plant diseases. The suppression of plant diseases has been extensively investigated in other organic amendments, such as manure and compost. Likewise, some studies have shown that vermicompost can suppress a wide range of microbial diseases, insect pests and plant parasitic nematodes. As regards the suppression of fungal diseases, the aqueous extracts of vermicompost are capable of reducing the growth of pathogenic fungi, such as *Botrytis cinerea, Sclerotinia sclerotiorum, Corticium rolfsii, Rhizoctonia solani and Fusarium oxysporum.*

Future Thrust

The technology of utilising forest and agricultural waste/biomass for the production of vermicompost needs to be popularised among foresters and ecologists. For this purpose, extensive training is required to be provided to the line departments of forestry and horticulture. The establishment of demonstration units (5 – 6 in no.) in every range of the division will definitely help in popularising the technology among the different wings of the forest department.

Conclusion

The concept of vermicompost technology in the forest department has not received due attention during the past as compared to the agriculture sector for many reasons. The success of composting depends upon the fecundity of the earthworm. It has the efficiency to consume all types of organic- rich waste material including leaf litter, vegetable waste, industrial waste, dairy farm waste, garden waste, sugar mill residues, slaughter house waste, hatcher waste and municipal waste. The protocol for the production of vermicompost from different sources of raw material has been developed recently by the Division of Basic Sciences and Humanities, Faculty of Agriculture, SKUAST-Kashmir. Three exotic earthworm species have been identified for efficient vermicomposting, but out of these three species, *Eisenia foetida* has shown an outstanding performance under temperate conditions in Kashmir Himalayas. The forest department of J&K has a tremendous scope to adopt the technology for raising quality nursery stock and rearing of vermiculture could also prove to be an additional source of income for the department.

References

Aalok, A., Tripathi A.K and Soni. P. 2008. Vermicomposting: A better solution for organic solid waste management. Journal of Human Ecology. 24, 59- 64

Bernal MP, Faredes C, Sanchez-Monedero MA, Cegarra J. 1998. Maturity and stability parameters of composts prepared with a, wide range of organic wastes. BioresourTechnol, 63:91–99

Bhadauria, T., and Ramakrishnan, P.S. 1996. Population Dynamics of earthworms and their activity in forest ecosystems of North- East. Journal of Tropical Ecology.7: 305- 318.

Brown G. G, Barois I, Lavelle P .2000. Regulation of soil organic matter dynamics and microbial activity in the drilosphere and the role of interactions with other edaphic functional domains. Eur J Soil Biol, 36:177–198

Bulluck, L.R., Brosius, M., Evanylo, G.K. and Ristaino, J.B. 2002. Organic and synthetic fertility amendments influence soil microbial, physical and chemical properties on organic and conventional farms. Applied Soil Ecology, 19, 147-160.

Cavani, N. and Mimmo, T. 2007. Rhizo deposition of Zea mays L. as affected by heterosis. Archives of Agronomy and Soil Science, 53, 593 – 604.

Chaoui, H.I., Zibilske, L.M. and Ohno, T. 2003. Effects of earthworm casts and compost on soil microbial activity and plant nutrient availability. Soil Biology and Biochemistry,35, 295-302.

Darwin F, Seward AC . 1903. More letters of Charles Darwin. In: John M (ed) A record of his work in series of hitherto unpublished letters, vol2., London, p 508.

Domínguez, J., Aira, M. and G6mez Brand6n, M. 2010. Vermicomposting: earthworms enhance the work of microbes. In: H. Insam, I. Franke-Whittle and M. Goberna, (Eds.), Microbes at Work: From Wastes to Resources (pp. 93-114).

Domínguez, J. 2004. State of the art and new perspectives on vermicomposting research.In: C.A. Edwards (Ed.). Earthworm Ecology (2nd edition). CRC Press LLC. Pp.401-424.

Gajalakshmi, S., and Abbasi, S.A.2003. High rate vermicomposting systems for recycling paper waste. Indian Journal of Biotechnology. 2: 613- 615.

Garcia-Gil, J.C., Plaza, C., Soler-Rovira, P. and Polo, A. 2000. Long-term effects of municipal solid waste compost application on soil enzyme activities and microbial biomass. Soil Biology and Biochemistry, 32 (13), 1907-1913.

Ghosh M, Chattopadhyay GN, Baral K .1999. Transformation of phosphorus during vermicomposting. BioresourTechnol 69:149–154

Lavelle E, Barois I, Martin A, Zaidi Z, Schaefer R .1989. Management of earthworm populations in agro-ecosystems: A possible way to maintain soil quality? In: Clarholm M, Bergstrom L (eds) Ecology of Arable Land: Perspectives and Challenges. Kluwer Academic Publishers, London, pp 109–122.

Lazcano, C., Sampedro, L., Zas, R. and Domínguez, J., 2010b. Assessment of plant growth promotion by vermicompost in different progenies of maritime pine (Pinuspinaster Ait.). Compost Science and Utilization ; 18, 111-118.

Lazcano, C., Sampedro, L., Zas, R. and Domínguez, J. 2010a. Vermicompost enhances germination of the maritime pine (Pinuspinaster Ait.). New Forest. 39, 387-400.

Prabakaran, J., 2005. Biomass resources in vermicomposting, In: proceedings of the State Level Symposium on Vermicomposting Technology for Rural Development, Madurai, Tamil Nadu, India. 27- 40.

Tognetti, C., Laos F., Mazzarino, M.J. and Hernández, M.T. 2005. Composting vs. vermicomposting: A comparison of end product quality. Compost Science and Utilization, 13, 6-13.

Weber, J., Karczewska, A., Drozd, J., Licznar, M., Licznar, S., Jamroz, E. and Kocowicz, A. 2007. Agricultural and ecological aspects of a sandy soil as affected by the application of municipal solid waste composts. Soil Biology and Biochemistry, 39,1294-1302.

10

Development of Nursery Technology of Chilgoza Pine (*Pinus gerardiana* Wall.): The Champion of Rocky Mountains

A.R. Malik, G.M. Bhat, P.A. Sofi, Nazir A. Pala, Ishtiyak Ahmad Peerzada and Aafaq A. Parrey

Faculty of Forestry Benhama Ganderbal 191 121, SKUAST-K, Jammu and Kashmir India

Conifers are valuable natural resources of the country, which contribute substantially to its socio-economic development by providing goods and services to the people and industry. They generate considerable revenue and also play a major role in enhancing the quality of the environment by influencing the basic life support system. Unfortunately, however, they have been subjected to over- exploitation and biotic pressure in the last many years, thereby threatening the very existence of some species such as yew, chilgoza pine etc. The genus *Pinus* is, in fact, one of the largest and most divergent of the coniferous genera, comprising about 95 species and numerous other varieties and hybrids. They are distributed widely in the northern hemisphere, chiefly in the temperate and cold climates. However, the distribution of *Pinus gerardiana* - the chilgoza or neoza pine - is very sparse. It is confined to the mountains of eastern Afghanistan, Pakistan, India and other scattered regions in the Hindu Kush Himalayas (latitude 30°N to 37°N and longitude 66°E and 80°E). In India, the chilgoza pine is restricted to the inner dry valleys of the North-west Himalayas, where it grows between 1600-3300 m above sea level. The tree is called Jhalgoza in Afghanistan, Chiri or chilgoza in Kashmir, Rhi or neoza in Kinnuar, Chujin in Chitral and Mirri in Pangi. In Himachal Pradesh, it occupies about 2060 hectares, with most of the area falling in Kinnaur district (2040 ha) and a small portion (20 ha) in Chamba district (Troup, 1921), extending westwards to Kishtwar and Astor in Jammu and Kashmir (Dogra, 1964; Critchfield and Little, 1966). *Pinus gerardiana* Wall., named after its discoverer, Captain Gerard, is a small to medium- sized evergreen tree, which varies between 8 m to 20 m in height and 1.5 to 3.5 m in girth. The wood of the tree is comparatively heavy amongst all the pines. (Plate 1). Its branches are short and horizontal forming a compact habit, while its bark is thin, smooth, glabrous, silver grey, having a mottled appearance and often, exfoliating in irregular, thin scales. The young shoots are olive in colour. The terminal buds are deep brown, non-resinous, ovoid and pointed at the tip. They vary from 1.0 to 1.5 cm in length and 0.21 to 0.4 cm in diameter (Kapoor *et al..*, 2003). The needles are 4.0 to 11.3 cm long, dark green in colour, densely arranged with three needles in each spur and are quite prominent in first year and fall in the second year. The cones are oblong, ovoid and glaucous when mature, while the scales are thick, woody and reflexed (Gamble, 1902). Flowering occurs from the last week of April up to the first week of June and pollination takes place thereafter. The male and female cones are borne on the same tree. The male cones are 0.6 to 1.5 cm long, cylindrical, reddish in colour and are born in clusters on young shoots. The female cones are terminal, erect and become pendulous after pollination. During the first year, the cones show a slight increase in growth, whereas during the second year, a rapid change occurs in their size and they become fleshy and attain full size by July. By the end of September/October of the same year, they are fully ripened and harvested. The mature cone normally varies from 10.5 to 26.0 cm in length and 6.5 to 12.5cm in diameter and possesses 65 to 98 thick and woody ovuliferous scales. The seeds are cylindrical and pointed at one end. They vary from 1.2 to 2.9 cm in length and 0.3 to 0.9 cm in breadth. Ectotrophic mycorrhizae have been reported from the roots of the trees. The rootlets appear slightly swollen and hairy towards the tips due to the setal hypae of the fungus mantle (Kapoor *et al.*, 2003). Its light demander nature and well- developed root system make it wind- firm. The species is also an excellent soil binder and prevents large-scale soil erosion from the otherwise loose and fragile strata of the region. The species occurs in association with *Cedrus deodara, Pinus wallichiana, Quercus ilex, Celtis australis, Daphne oleoides, Artemisia maritime, Rosa webbiana, Berberis spp, Desmodium spp, Indigofera gerardiana, etc.*

Regeneration

Natural: The area under chilgoza forest has already shrunk to about 2000 hectares in Himachal Pradesh because each and every cone is lopped, leaving very little for natural regeneration (Tandon, 1963; Singh *et al.,*1973) (Plate 1). The species is thus facing extinction and has been included in the list of endangered species (Sehgal and Sharma, 1989). (However, the chilgoza forest of Kishtwar area is not studies by ant research so far). ??Today, the overall natural regeneration status of the species is very poor (15%). Thus, the species is facing higher risk of extinction and needs to be considered as 'Critically Endangered' in the Indian Himalayan region (Malik *et al.*, 2012).

Artificial: In view of the limited distribution and failure of natural regeneration, the necessary measures were undertaken for artificial regeneration. The chilgoza pine seeds are also known to possess both physiological and morphological dormancy (Sharma, 2005 and Malik, 2007). The outdoor (Pit) stratification of 45-60 days was employed for significantly accelerated germination in the species under laboratory conditions. However, under nursery conditions, 60 days stratification at 4$\pm$1oC followed by soaking in 400 ppm GA_3 produced the significantly best overall seedling growth performance in the species. (Malik *et al.,* 2008; Malik and Shamet, 2008).

The storage of seeds assumes importance in view of infrequent and erratic seed years to provide sustained supply of seeds for afforestation programmes in the species. For this, the chilgoza pine seeds were stored to check the their viability and longevity. The seeds stored in earthen pots at 0$\pm$1oC performed significantly better by achieving highest germinability after one year of storage (Malik *et al.*, 2013).

The raising of chilgoza seedlings in nurseries was standardised by using different containers and potting media. The bottom- perforated polybags (23x10cm), filled with a mixture of chilgoza forest soil, sand, moss and farm yard manure (1:1:1:1) proved superior, with respect to various seedling growth parameters in chilgoza pine under nursery conditions (Malik *et al.,* 2009). The large -sized seeds (>2.46 cm) registered better seedling growth as compared to small and medium- sized seeds in the species. The seedlings attained average size (68.5 cm) after two years of growth (Malik, 2007). The other standard nursery practices should be followed accordingly for maintaining the growth and development of the seedlings (Plate 1).

Economic and Ecological Significance

Dry fruit: Chilgoza pine is the only conifer which provides edible kernels/ nuts available in India. The fresh seeds are rich in sugar (4.07%), proteins (13.03%), oils (52.15%) and moisture (25.36%) (Malik, 2007). The species plays an important role in the socio-economic upliftment of the people in the tribal area of Himachal Pradesh and Jammu and Kashmir (Sehgal and Khosla, 1986).

Wood: The chilgoza pine is sometimes used as timber and traditional torch wood during nights.

Resin: The chilgoza pine is an excellent source of fine turpentine resin.

Medicine: The seeds are used as anodyne and stimulants. Their oil is also used against wounds and ulcers as it has healing properties.

Food: The local population used chilgoza nuts/ kernels mixed with floor as dietable food.

Plate 1: (A) Pinus gerardiana: anthropogenic pressure on mature individuals; **(B)** Mature seeds; **(C)** Seed germination and **(D)** Seedling growth

Forest ecology: The indigenous nature of the species has the capacity to withstand the extremes of cold climates and aridity. The species is the best soil binder and grows in the excessively dry, bare, rocky and barren hills, under difficult site conditions prevailing in the Himalayas. Due to its peculiar characteristics, chilgoza pine can be described as the **"Champion of the Rocky Mountains".**

Market trade: Due to the high marketable value and demand of chilgoza nuts, the right holders collect almost each and every cone from the natural forest and sell them for high rates ranging from Rs.350-650/kg, depending upon the demand and supply position. A survey of the market conducted by the Forest and Revenue departments of District Kinnuar revealed that roughly 80-120 tons of chilgoza nuts are collected by right holders and exported to the plains every year from Kinnaur district alone (Personal survey, 2004-06). The market channel from collection to end point is shown in Figure 1.

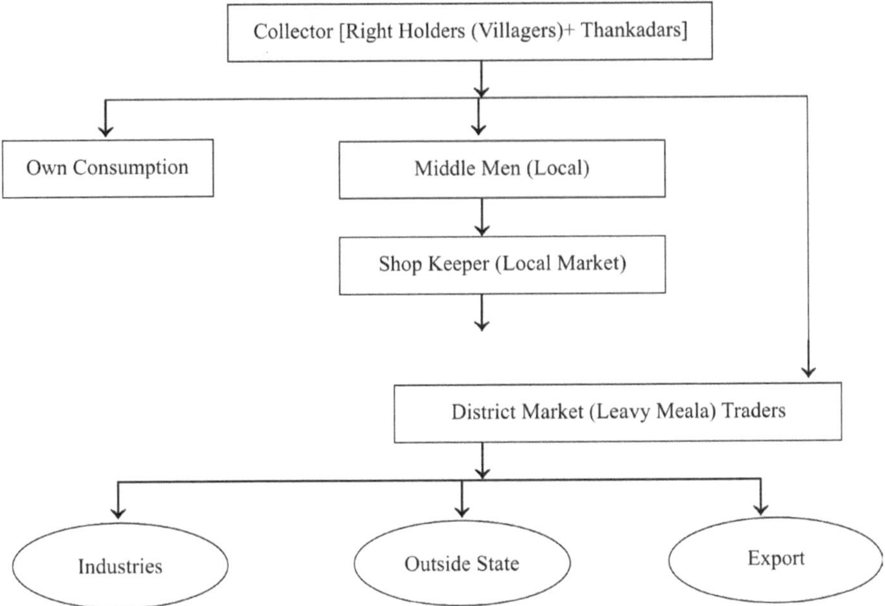

Fig. Market channel and people involved in chilgoza nut trade in Kinnuar.

References

Critchfield, W. B. and E. L. Little, 1966. Geographic distribution of the Pinus in the world. USDA Forest Service pp 11-15.

Dogra, P. D. 1964. Gymnosperms of India–II. Chilgoza pine (Pinus gerardiana Wall.). Bulletin of the National Botanic Garden No. (109): 446 p.

Gamble, J. S. 1902. A Manual of Indian Timbers. Humphrey Milford, Oxford University Press.709 p.

Kapoor, K. S., Kimar Surinder and Singh Ombir, 2003. Chilgoza (Pinus gerardiana Wall):champion of the rocky Mountains. HFRI Bro. No.8.Pp16.

Malik, A.R., 2007. Studies on natural regeneration status and nursery technology in Chilgoza pine (Pinus gerardiana Wall.), Ph.D. Thesis. Dr. YS Parmar University of Horticulture and Forestry, Nauni, Solan, India., p. 216.

Malik, A.R., G.S. Shamet, J.S.Butola, G. M.Bhat, A.A.Mir and Gowher Nabi. 2013. Standardization of seed storage conditions in Chilgoza pine (Pinus gerardiana Wall.):an Endangered pine of Hind Kush Himalaya. Tree -Structure and Function-DOI10.1007/ s00468-013-0868-y

Malik, A. R., G.S. Shamet And J.S.Butola. 2012. Natural regeneration status of chilgoza pine (Pinus gerardiana wall.) In Himachal Pradesh, India: An endangered pine of high edible value. Applied Ecology and Environmental Research 10(3): 365-373

Malik, A. R., G. S. Shamet and M. Ali, 2008. Seed stratification of Pinus gerardiana: Effect of stratification duration and temperature. Indian Forester. 134 (8): 1072-1078.

Malik, A. R., G. S. Shamet and M. Ali, 2009. Germination and seedling growth of Pinus gerardiana in nursery: Effect of stratification period and temperature. Indian Journal of Forestry. 32 (2):

Malik, A. R. and G. S. Shamet, 2008. Germination and biochemical changes in the seeds of chilgoza pine (Pinus gerardiana Wall.) by stratification: an endangered conifer species of north-west Himalaya. Indian Journal of Plant Physiology. 13 (3): 278-283.

Sehgal, R. N. and P. K. Khosla, 1986. Chilgoza pine the threaded social forestry tree of dry temperate Himalaya. National Symposium on Research in Social Forestry for Rural Development. January 1-2, 1986.

Sehgal, R. N. and P. K. Sharma, 1989. Chilgoza the Endangered Social Forestry Pine of Kinnaur. Dr. YS Parmar University of Horticulture and Forestry, Nauni-Solan, India. Technical Bulletin No FBTI (1): pp 1-8.

Sharma, Veena. 2005. Effect of artificial stratification of Chilgoza pine (Pinus gerardiana) seeds on its germination. In: Short Rotation Forestry for Industrial and Rural Development. eds: K S Verma, D K Khurana and Lars Christarsson. 2005 ISTA, Nauni, Solan, India pp 266-270.

Singh, R. V., D. C. Khanduri and Lal. Kashmiri, 1973. Chilgoza pine (Pinus gerardiana) regeneration in Himachal Pradesh. Indian Forester. 3: 126-133.

Tandon, J. C. 1963. Revised Working Plan for the Kinnaur and Kochi Forests (Upper Sutlej valley). Himachal Pradesh 1961-62 to 1975-76.

Troup, R. S. 1921. Silviculture of Indian Trees. Vol. III. Clarendon Press, Oxford University.1090-1093

11

Raising of Multipurpose Broad-leaved Tree Species in Nursery

Vaishnu Dutt[1], A.H. Mughal [2], Shiba Zahoor[1] and Nazir A. Pala[2]

[1]*Faculty of Forestry, SKUAST-K, Benhama-Watlar, Ganderbal, Jammu and Kashmir, India*

[2]*Associate Director Research SKUAST-K, Jummu and Kashmir, India*

Introduction

The term 'multipurpose trees' covers all woody perennials used in agroforestry. Many authors have given many definitions of MPTs and MPTS. Burley and von Carlowitz (1984) reviewed many definitions and gave a synthesised definition which states that "multipurpose trees and shrubs (MPTS) are those which are deliberately grown and kept and managed for preferably more than one intended use, usually economically and/or ecologically motivated major products and / or services in any multipurpose land use systems, especially agroforestry systems". The Forestry /Fuelwood Research Development Project (1992) precisely defined 'multipurpose trees' (MPTs) to mean tree species that are grown to provide more than one significant crop or function on the farm. For instance, on small farms, this can often mean that the farmer uses both wood and leaves from the same tree. Any woody perennial species can be 'multipurpose' in one situation and 'single purpose' in another.

There are large numbers of MPTs belonging to a wide taxonomic range. Farmers have been growing trees for different purposes for thousands of years. Certainly, all trees provide shade and protection from soil erosion. In this sense, all trees can be said to have at least two purposes. More precisely, the term 'multipurpose tree species' denote tree species that are grown to provide more than one significant crop or function or form. These may include fuelwood, timber, fibre, fodder, food, medicine and environmental services such as soil conservation and fertility improvement, climatic amelioration, flood control, carbon sequestration etc. Tree species can be multipurpose in two ways:

- **A single tree can yield more than one crop.** For example, farmers in Central America grow *Gliricedia sepium* as living fences that provide fuelwood, fodder and green manure for agricultural crops, all at the same time. Similarly, farmers in some tropical regions grow jackfruit (*Artocarpus heterophyllus*) that provides food, fodder and timber.

- **Trees of the same species, when managed differently, can yield different crops.** In tropics, for example, *Leucaena leucocephala* is so managed that some trees will principally yield wood, while others will principally produce leaf fodder.

Farmers can grow multipurpose trees in various combinations with other crops, as in agroforestry, block plantation of trees or naturally regenerating tree farms. In certain settings, multipurpose trees are grown and managed for only one purpose. In the Solomon Islands, *Gliricidia sepium* is grown only to provide shade in coffee plantations. The same species is planted in Central America and the Philippines for multiple uses, yet the way it is managed results in a very different shape and use. Farmers facing changing conditions in their environment or market can also change the way they manage a tree. For example, market changes may persuade farmers, who previously grew trees for fodder and fuel, to cut their trees for sale as round wood for construction material (Antony and Lal, 2014).

Multipurpose trees are trees that are deliberately grown and managed for more than one output. They may supply food in the form of fruit, nuts or leaves that can be used as a vegetable. At the same time, they may supply firewood, add nitrogen to the soil or supply some other combination of multiple outputs. 'Multipurpose tree' is a common term in agroforestry, particularly when speaking of tropical agroforestry, where the tree owner is a subsistence farmer.

While all trees can be said to serve several purposes, such as providing habitat, shade or soil improvement, multipurpose trees have a greater impact on a farmer's well- being because they fulfill more than one basic human need. In most cases, multipurpose trees have a primary role, such as being part of a living fence or a windbreak or being used in an ally cropping system. In addition to this, they will have one or more secondary roles, most often supplying a family with food or firewood or both.

When a multipurpose tree is planted, a number of needs and functions can be fulfilled at once. They may be used as a windbreak, while also supplying a staple food for the owner. They may be used as a fence post in a living fence, while also being the main source of firewood for the owner. They may be intercropped into existing fields to supply nitrogen to the soil and at the same time, serve as a source of both food and firewood.

Common Multipurpose Trees of the Tropics Include

- *Gliricidia sepium*: It is the most common tree used for living fences in Central America. It provides firewood and fodder and supplies nitrogen to the soil.
- Moringa (*Moringa oleifera*): It provides edible leaves, pods and beans, commonly used for animal forage and shade. However, it does not fix nitrogen as is commonly believed.
- Neem – It has limited use as an insect repellent and antibiotic. It adds nitrogen to the soil and is used in windbreaks. Its biomass production is used as mulch and firewood.

  Ideally, most trees found on tropical farms should be multipurpose and provide more to the farmer than simply shade and firewood. In most cases, they should be nitrogen fixing legumes or trees that greatly enhance the farmer's food security.

Nursery Raising of Multipurpose Broad-leaved Tree Species

A nursery is a place where plants are cultivated and grown to usable size. Nursery techniques involve raising seedlings and saplings as well as grafting economically useful and ornamental plants through scientific methods. Several new techniques are available, which are cheap and effective. These new techniques are useful in increasing the success rate of grafting and rooting of cuttings, thereby increasing seedling vigour, reducing transplanting shock and also, the overall quantum of manual work. Nursery management gained the status of a commercial venture where retailer nurseries sell planting materials to the general public and nurseries, which sell only to other nurseries and to commercial landscape gardeners and private nurseries, which supply the needs of the institutions of private estates.

1. The sites for the nursery should be selected at such places where abundant sunshine and proper ventilation is available.
2. The nursery site should be on a higher location so that water stagnation can be avoided.
3. In humid and rain- prone areas, the nursery should be well- protected from heavy rains by protective structures.
4. The site should be nearer to irrigation facilities and accessible.
5. The site should be protected from stray animals, snails, rats etc.
6. The soil should be sandy loam or loamy with a pH range of 6 to 7, rich in organic matter and free from pathogenic inocula.

Necessity of Nursery

A nursery is a necessity for every grower. The development of seedlings in nurseries not only reduces the crop span, but also increases the uniformity of the crop and thus, harvesting, as compared to direct- sown crops. The transplanting of seedlings also eliminates the need of thinning and provides good opportunities for virus- free, vigorous off- season nurseries, if grown under protected conditions. A nursery is helpful and convenient to manage seedlings under a small area and the grower can obtain timely plant protection measures with minimal efforts. A nursery provides a favourable climate to emerging plants for their better growth and development. Nursery production helps in maintaining effective plant stands in the shortest possible time through gap filling.

Although seeds can be sown directly in the field, experiments have shown that raising seedlings in a nursery has a number of advantages as discussed below.

1. Intensive care: Seedlings receive better care and protection in the nursery. The average garden soil is not an ideal medium for raising seedlings, especially from the point of view of soil tilth. At an early stage of development, most vegetable crops require special attention that is not possible in the main field.
2. Reduction of costs: Fewer seeds are used for raising seedlings in the nursery than for sowing directly in the field, because in the latter, seedlings have to be thinned, which is wasteful. Therefore, when expensive hybrid seeds are used, transplanting becomes economically more attractive. The cost of pesticides and labour is also reduced under nursery conditions, as compared to planting directly in the field.
3. Opportunity for selection: Raising seedlings in a nursery afford the grower an opportunity to select well-grown, vigorous, uniform and disease- free seedlings.
4. Extend a short growing season for late maturing crops: Seedlings can be raised in a nursery in a protected environment before conditions outside become suitable for growth and can be transplanted into the field when conditions allow, thus reducing the amount of time spent in the field.

Nursery Management

The first stage in the successful production of tree/horticulture crops is to raise healthy vigorous seedlings. The seeds of tree crops are raised in nurseries and the prepared seedlings are then planted in the field. Young plants, whether propagated from seed or vegetative parts, require a lot of care, particularly during the early stages of growth. They have to be protected from adverse temperature, heavy rain, wind and a variety of pests and diseases. A tree or fruit nursery is a place where plants are taken care of during the early stages of growth, providing optimum conditions for germination and subsequent growth, until they are strong enough to be planted out in their permanent place. A nursery can be as simple as a raised bed in an open field or as sophisticated as a glasshouse with micro-sprinklers, but the management and plant propagation are the same. and interrelated. In fact, the mass multiplication of quality planting material is the central theme of nursery management. Nursery management is a team effort to reach the desired goal.

The main phases of nursery management are (1) Planning- edaphic, climatic and socio-economic considerations; demand for planting materials; provision of mother block, requirement of land area, water supply, working tools, growing structures and input availability; accessibility; trained man power; plant protection; disposal of planting materials etc. (2) Implementation- land treatment, protection against biotic interference and soil erosion, proper layout, input supply (3) Monitoring- physical presence, rapid response, critical analysis, intensity of work and (4) feedback for further improvement. The key elements of hi-tech nursery management are the place, plant and the person behind it.

Advantages of Nursery Management

- It is possible to provide favourable germination and growth conditions.
- It provides better care for younger plants as it is easy to protect them against pathogenic infection, pests and weeds in the small area of a nursery.
- The crop grown by nursery raising is quite early and fetches a higher price in the market. Therefore, it is economically more profitable.
- There is saving of land and labour as crop rotation can be followed in the main fields.
- More time is available for the preparation of the main field because seedlings are grown separately in the nursery.
- In the case of expensive seeds, we can economise by sowing them in the nursery.

Layout of Model Nursery

A nursery is the place where all kinds of plants like trees, shrubs, climbers etc. are grown and kept for transplanting or for using them as stock plants for budding, grafting and other methods of propagation or for sale. The modern nurseries also serve as areas where garden tools and fertilisers are offered for sale along with plant material.

1. ***Fence:*** Prior to the establishment of a nursery, a good fence with barbed wire must be erected all around it to protect the plants from animals and theft. The fence could be further strengthened by planting a live hedge with thorny fruit plants. This not only adds beauty in bearing, but also provides additional income through the sale of fruits and seedlings obtained from the seeds.
2. ***Roads and paths:*** A proper planning of roads and paths inside the nursery will not only add beauty, but also make the nursery operation easy and economical. This can be achieved by dividing the nursery into different blocks and various sections.
3. ***Progeny block / mother plant block:*** The nursery should have a well-maintained progeny block or mother plant block/scion bank planted with those varieties that are in demand. The layer cutting should be obtained preferably from the original breeder/research institute from where it is released or from a reputed nursery. One should remember that the success of any nursery largely depends upon the initial selection of progeny plants or mother plants for further multiplication. A well-managed progeny block or mother plant block will not only create confidence among the customers, but also reduce the cost of production and increase the success rate of grafting/budding/layering because of the availability of fresh scion material throughout the season within the nursery itself. Additionally, there will not be any lag period between the separation of scions and grafting age.
4. ***Irrigation systems:*** Nursery plants require an abundant supply of water for irrigation, since they are grown in poly bags or pots with a limited quantity of potting mixture. Hence, a sufficient number of wells are required in a nursery for providing adequate irrigation water. In areas with low water yields and frequent power failures, a sump to hold sufficient quantity of water to irrigate the nursery plants is also essential along with appropriate pumps for lifting the irrigation water. In areas where electricity failure is a common problem, an alternative power supply is essential for the smooth running of the pump set. Since water security is a limiting factor in most areas in the country, well laid out PVC pipeline systems will solve the problem to a greater extent. This facilitates an efficient and economic distribution of irrigation water to various components in the nursery.

5. ***Office-cum-stores:*** An office-cum-store is needed for the effective management of the nursery. The office building may be constructed in a place which offers better supervision. The office buildings may be decorated with attractive photographs of fruits. A store room of suitable size is needed for storing poly bags, tools and implements, packaging materials, labels, pesticides, fertilisers etc.
6. ***Seed beds:*** In a nursery, this component is essential to raise the seedlings and rootstock. These are to be laid out near the water source, since they require frequent watering and irrigation. Beds of 1.0 meter width of any convenient length are to be made. A working area of 60 cm between the beds is necessary. These features provide ease in sowing seeds, weeding, watering, spraying and lifting of seedlings. Irrigation channels are to be laid out conveniently. Alternatively, the sprinkler irrigation system may be provided for watering the beds, which aids uniform germination and seedling growth.
7. ***Nursery beds:*** Raising of seedling / rootstalk in poly bags requires more space compared to nursery beds, but mortality is greatly reduced along with uniformity. The nursery bed area should also have a provision to keep the grafted plants in trenches of 30 cm depth and 1.0 m width, so as to accommodate 500 grafts in each bed. Alternatively, the graft / layer can be arranged on the ground in beds of 1.0 m width with 60 cm working space in- between the beds. Such beds can be irrigated either with a hose fitted to a flexible hosepipe or by overhead micro-sprinklers.
8. ***Potting mixture and potting yard:*** For better success of nursery plants, a good potting mixture is necessary. The potting mixture may be prepared well in advance by adding sufficient quantity of super phosphate for better decomposition and solubilisation. The potting mixture can be kept near the potting yard, where potting is done.
9. ***Containers and Beds:*** The date of sowing should be scheduled so as to allow the time necessary for the species to attain a suitable size by the planned date of planting. Most fast-growing species take three to six months to reach plantable size, but each nursery must determine its own rate of development for each species.

Various containers for sowing are commonly used. Currently, the most popular container is a black polyethylene bag. The bag is perforated with 8 to 12 holes from the bottom up to half the total depth. Plastic bags are popular because they are relatively cheap and convenient to use. However, the round shape encourages undesirable coiling of the roots and the drainage holes, although essential, permit the roots to emerge and enter the underlying soil.

Beds and containers should be well-drained and usually kept above the mean ground line to prevent water logging, improve aeration and reduce root rot.

Seedling Care

Care of the seedlings (also called culture) is necessary from the time the seed is sown until the planting stock is dispatched to the field. Seedlings are the most delicate during the first three months after germination.

The techniques of sowing, mode of propagation and uses of multipurpose trees and shrubs commonly found in India are discussed in Table 1.

Table 1. Techniques of sowing, Mode of propagation and uses of multipurpose trees and shrubs commonly found in India.

S. No	Botanical/ common names	Family	Time of sowing	Mode of propagation	Uses
1.	*Acacia Arabica* (Babul)	Mimosaceae	May-June	Seed	Fodder, soil conservation, agricultural implements, medicine, tannin, fuel
2.	*Acacia auriculiformis* (Babul)	-do-	April-May	Seed	Fodder, tannin, fuel, Nitrogen fixing, Soil amelioration, timber
3.	*Acacia catechu* (Khair)	-do-	May-June	Seed	Fodder, fuel, poles, soil conservation & katha etc.
4.	*Aegle marmelos* (Bel)	Rutaceae	May-June	Seed, sucker	Fuel, fruits (bark & gum used as medicine)
5.	*Ailanthus excelsa* (Maharukh)	Simarubaceae	July	seed, cutting	Fodder, fuel, soil amelioration, timber
6.	*Albizia chinensis* (Kala siris)	Mimosaceae	July-August	Seed	Timber, fodder, fuel, poles, agricultural implements
7.	*Albizia lebbek* (Siris)	Mimosaceae	July-August	Seed	Timber, fodder, fuel, agricultural implements,
8.	Albizia procera (Safed Siris)	Mimosaceae	July-August	Seed	Timber, fodder, fuel, poles, medicinal
9.	*Anacardium occidentale* (Kaju)	Anacardia-ceae	June	Seed	Fruit, fuel
10.	*Anthocephalus chinensis* (Kadam)	Rubiaceae	March-May	Seed	Fodder, fuel, pulpwood, ornamental
11.	*Artocarpus heterophyllus* (Kathal)	Moraceae	July-Aug	Seed	Fruit, ornamental, timber, fuel (leaves, roots and flowers used as medicine)

12.	*Artocarpus lakoocha* (Barhar)	Moraceae	July	Seed	Fruit, timber, fuel
13.	*Azadirachta indica* (Neem)	Meliaceae	July	Seed	Planks,toothbrush, medicine (bark, leaves, gum)
14.	*Bauhinia variegata* (Kachnar)	Caesalpinia-ceae	May	Seed	Fodder, food and timber; leaves and flowers used as vegetables, medicines
15.	*Cassia fistula* (Amaltas)	Caesalpinia-ceae	March-April	Seed	Medicinal and fuel, ornamental
16.	*Casuarina equisetifolia* (Chok)	MImosaceae	Feb-Nov	Seed	Fuel, timber good in coastal area for plantation
17.	*Dalbergia latifolia* (Sissoo)	Papilionaceae	Feb-March	Seed	Fodder, timber, fuel
18.	*Dalbergia sissoo* (Shisham)	Papilionaceae	March-April	Seed	Timber, fodder, fuel, soil conservation
19.	*Emblica officinalis* (Amla)	Euphorbia-ceae	May-June	Seed	Fruit, fuel, ornamental, medicinal
20.	*Eucalyptus camandulensis* (Eucalyptus)	Myrtaceae	Sept-Oct	Seed	Gum, oil, tannin, ornamental, soil conservation, industrial timber
21.	*Eucalyptus globulus* (Blue gum)	Myrtaceae	Sept-Oct	Seed	Pulpwood, fuel, timber
22.	*Eucalyptus teriticornis* (Mysore gum)	Myrtaceae	Sept-Oct	Seed	Fuel and Pulpwood
23.	*Ficus religiosa* (Pipal)	Moraceae	March-May	Seed	Avenue trees, fodder, fuel fruit, medicinal
24.	*Gmelina arborea* (Gamhar)	Verbenacea	May-June	Seed	Furniture wood
25.	*Grevillea robusta* (Silver oak)	Proteaceae	June	Seed	Ornamental, avenue tree
26.	*Grewia spp.* (Dhaman)	Tiliaceae	April	Seed	Fuel and fodder
27.	*Hardwickia binata* (Anjan)	Caesalpinia-ceae	May-June	Seed	Fodder, fuel, agricultural implements
28.	*Juglans regia* (Walnut)	Caesalpinia-ceae	Jan-Feb	Seed	Fruit, carving and ornamental (timber, gum, butter)

29.	*Leucaena leucocephala* (Su-babul)	Mimosaceae	June	Seed	Poles, fodder, fuel, soil conservation
30.	*Madhuca latifolia* (Mahua)	Sapotaceae	July-Aug	Seed	Flower, fruit, timber, fodder, oil (bark seeds, flower as medicine)
31.	*Mangifera indica* (Am)	Anacardia-ceae	June-July	Seed	Fruit, timber, fodder, medicinal (seeds, leaves, bark)
32.	*Melia azedirach* (Bakain)	Meliaceae	Feb-April	Seed	Avenue tree
33.	*Moringa oleifera* (Sahajan)	Meringaceae	Aril-June	Stem cutting, Seed	Fruits, fodder, medicinal (leaves and fruits)
34.	*Morus alba* (Mulberry)	Moraceae	April-May	Stem cutting, Seed	Fodder, fruit, sports good
35.	*Pongamia pinnata* (Karanj)	Papilionaceae	April-May	Seed	Flowers used as fertiliser, seeds for oil; fuel, soil conservation
36.	*Populus spp.* (Poplar)	Salicaceae	Feb-March	cutting, seed	Soil conservation, fodder, packing cases, pulpwood, fuel, matchbox and splints
37.	*Prosopis chilensis* (Kabuli)	Mimosaceae	May-June	Seed	Fodder, fuel, soil conservation
38.	*Prunus armeniaca* (Apricot)	Rosaeceae	Sept-Oct	Seed	Fruit, small timber, poles, ornamental, soil conservation
39.	*Psidium guajava* (Amrud)	Myrtaceae	March-April	Seed	Fruit, fodder, fuel
40.	*Quercus spp.* (Indian Oak)	Fagaceae	Dec-Feb	Seed	Fodder, fuel, silk cocoon host trees
41.	*Salix spp.* (Willow)	Salicaceae	Feb-March	Cutting, pollards	Fodder, fuel, silk cocoon host trees
42.	*Santalum album* (Sandalwood, Chandan)	Santalaceae	May-Aug; Oct-Dec	Seed, cutting	Industrial timber, oil, ornamental oil, medicinal, soil conservation
43.	*Sapindus mukorossi* (Ritha)	Sapindaceae	March-April	Seed	Fuel, soap, nut used in washing clothes
44.	*Sesbania grandiflora* (Basna)	Papilionaceae	June-July	Seed	Flowers, fence posts, tannin, fodder, fuel pulp, medicinal, fodder for milch cattle

45.	*Shorea robusta* (Sal)	Dipterocarpa-ceae	May-June	Seed	Structural and industrial timber, poles, fuel, tannin
46.	*Syzygium cumini* (Jamun)	Myrtaceae	June-July	Seed	Fruit, timber, fuel, medicinal (flowers, fruits)
47.	*Tamarindus indica* (Imli)	Caesalpina-ceae	April	Seed	Fruit, fuel, medicinal, timber, leaves, flowers as medicine
48.	*Tectona grandis* (Sag)	Verbenaceae	April-May	Seed	Structural and industrial timber, poles, fuel, tannin, ornamental, medicinal
49.	*Terminalia arjuna* (Arjun)	Combreta-ceae	Feb-May	Seed	Industrial timber, small poles, fuel, fodder, timber, ornamental,medicinal (bark)
50.	*Terminalia chebula* (Harde)	Combreta-ceae	Feb-May	Seed	Fruit, fuel, medicinal
51.	Terminalia *tomentosa* (Asain sain)	Combreta-ceae	Feb-May	Seed	Industrial timber, tannin, gum, medicinal, poles, fuel
52.	*Zizyphus jujuba* (Ber)	Rhamnaceae	Feb-April	Seed	Fruit, fodder, poles

Source: Chundawat and Gautam (1993)

There are many species of multipurpose trees and shrubs found in the Kashmir valley. The nursery practices and uses of some important broad-leaved multipurpose trees of the Kashmir valley are discussed below:

Nursery practices and uses of important broad-leaved multipurpose trees of the Kashmir valley

1	Scientific name Local name English name Propagation Uses	*Acer ceasium* Kinar Maple Seeds are sown from November to February in nursery. Fodder and fuel wood also used in turnery articles. The plant is also grown for ornamental purpose.
2.	Scientific name Local name English name Propagation Uses	*Aesculus indica* Handoon Horse chestnut Seeds are sown in November in nursery. Fuel wood and fodder.
3.	Scientific name Local name English name Propagation Uses	*Albizia julibrissin* Albizia Persian silk tree Seeds are sown from November to February in nursery. The plant is grown for fodder, fuel wood, bee forage and ornamental purpose.

4	Scientific name Local name English name Propagation Uses	*Ailanthus altissima* Alanthrus Tree of heaven Seeds are sown from November to February in nursery beds. Suckers are also used of propagation. Soil conservation, Ornamental, Wood flexible and used for making spoons and other kitchen utensils.
5.	Scientific name Local name English name Propagation Uses	*Alnus nitida* Champ kul Alder Seeds are sown from November to February in nursery beds Match sticks, dye yielding. furniture, cabinets, and other woodworking products
6.	Scientific name Local name English name Propagation Uses	*Betula utilis* Burza. Bhoj patri Birch Propagated through soft wood cuttings and seeds in March in nursery. Earlier used in place of paper. Used as Fuel wood, fodder, charcoal and bark is used for roofing.
7	Scientific name Local name English name Propagation Uses	*Castanea* *sativa* Gour Sweet Chestnut Seeds are sown from November to February in nursery. Best known for its edible nuts used by confectioners, eaten roasted and ground to make flour. Wood of the tree is durable and is used to make furniture, barrels, fencing and roof beam.
8.	Scientific name Local name English name Propagation Uses	*Celtis australis* Brimij Hackberry Seeds are sown in November December. Moist cold stratified seeds for one month perform better in nuresery Fuel wood and fodder
9	Scientific name Local name English name Propagation Uses	*Ficus palmate* Injir Fig Cuttings are sown from November to March in nursery Medicinal use, ornamental and for fruits
10	Scientific name Local name English name Propagation Uses	*Fraxinus floribunda* Hom Ash Seeds are sown in nursery in the month of November after cold stratification for 4 weeks Timber for boats and agricultural implements

11	Scientific name Local name English name Propagation Uses	*Juglans regia* Doon Walnut Seeds are sown from November to December in nursery beds Nuts are edible and its wood used for furniture and fuel wood.
12	Scientific name Local name English name Propagation Uses	*Melia azaderach* Derk Pertian lilac Seeds are sown from November to February in nursery. Fuel wood, soil conservation.
13	Scientific name Local name English name Propagation Uses	*Morus alba* Safed Tul Mulberry Seeds are sown from November to February in nursery beds and through cutting in February-March. Timber for furniture and sericulture
14	Scientific name Local name English name Propagation Uses	*Parrotia jacquemontiana* Posh Parrotia Suckers are sown from November to February in nursery beds Agricultural implements, wicker work and tool handles.
15	Scientific name Local name English name Propagation Uses	*Plaltinus orientalis* Boin Plane tree Shoots, branch cuttings are sown from November to February in nursery beds. Aesthetic purpose, timber, fuel wood.
16	Scientific name Local name English name Propagation Uses	*Populus alba* Duda phrast Poplar Branch cuttings are sown from November to February in nursery beds. Timber, plywood veneer and fodder.
17	Scientific name Local name English name Propagation Uses	*Populus ciliata* Pahari phrast Poplar Branch cuttings are sown from November to February in nursery beds. Timber, plywood veneer and fodder.
18.	Scientific name Local name English name Propagation Uses	*Populus deltoides* Punjabi phrast Poplar Branch cuttings are sown from November to February in nursery beds. Timber, plywood veneer and fodder.

19	Scientific name Local name English name Propagation Uses	*Populus nigra* Pahari phrast Poplar Branch cuttings are sown from November to February in nursery beds. Timber, plywood veneer and fodder.
20	Scientific name Local name English name Propagation Uses	*Prunus armeniaca* Teth czer Wild apricot Seeds are sown in November or December in the nursery. Fruit, oil, fodder, charcoal and agricultural implements.
21	Scientific name Local name English name Propagation Uses	*Punica* *granatum* *Den/ Annar* Pomegranate Shoots are planted from November to February in nursery beds. Shoots are planted from November to February in nursery beds. Fruit, medicinal use leather tanning, fodder, aesthetic and charcoal.
22	Scientific name Local name English name Propagation Uses	*Quercus ilex* Jamal goat Oak Seeds are sown from November to February in nursery beds. Fuel wood, charcoal and soil conservation.
23	Scientific name Local name English name Propagation Uses	*Quercus robur* Hoom Oak Seeds are sown from November to February in nursery beds and can also propagated through softwood and hard wood cuttings. Fuel wood, charcoal and soil conservation.
24	Scientific name Local name English name Propagation Uses	*Robiana pseudoacacia* Kiker Black locust Seeds are sown from November to February after immersing in boiling water in nursery beds. Fodder, fencing poles, fuel wood, charcoal and soil conservation.
25	Scientific name Local name English name Propagation Uses	*Salix alba var. caerulea* Bat willow Willow Through cuttings in nursery. Sports goods, fodder, fuel wood and charcoal .
26	Scientific name Local name English name Propagation Uses	*Salix babylonica* Chateri veer Weeping willow Through cuttings in nursery in November to March. Packing boxes, sports goods, ornamental purpose, paper pulp, fodder, fuel wood and charcoal .

27	Scientific name Local name English name Propagation Uses	*Salix caprea* Bed mushk Goat willow Through cuttings in nursery in November to March. Perfumery, paper pulp, fodder, fuel wood and charcoal.
28	Scientific name Local name English name Propagation Uses	*Salix matsudana* Ring veer Cock and screw willow Through stem cuttings in nursery. Aesthetic
29	Scientific name Local name English name Propagation Uses	*Salix trianda* Kani veer Curly willow Through cuttings in nursery Asthetic

Source: Mughal and Mugloo, 2015

4. Multipurpose Broad-leaved Tree Species Identified for Agroforestry in the Kashmir Valley

The following 12 multipurpose tree species have been identified as the most promising with respect to their potential in agroforestry systems (Mughal, 2017).

Species	Method of planting	Uses				
		Fruit	Fodder	Fuel wood	Timber	Soil conser-vation
Aesculus indica	D.S	-	+++	++++	+++++	++
Ailanthus altissima	D.S	-	-	+	+++	+++++
Juglans regia	D.S	+++++	-	+++++	+++++	+++
Morus spp.	E.P	+++++	+++	+++	+++	++++
Populus spp.	B.P	-	+++	+++++	++++	+++
Prunus armeniaca	D.S, E.P	+++++	+++	+++++	+++	+++++
Prunus cerasus	D.S, E.P	+++++	+++	+++++	+++	++++
Prunus persica	D.S, E.P	+++++	+++	+++++	+++	++++
Pyrus communis	E.P	+++++	+++	+++++	+++	++++
Robinia pseudoacacia	R.S	-	+++++	+++++	+++	+++++
Salix spp.	B.P	-	+++++	++++	++++	+
Ulmus villosa	E.P	-	+++++	+++++	+++	++++

D.S- Direct sowing E.P- Entire planting B.P- Branch/cutting planting R.S- Root sucker	- No use ++ Satisfactory +++ Good ++++ Better +++++ Best

Some work in the area of growing multipurpose broad- leaved tree species in nurseries has been done in the Faculty of Forestry.

1. *Morus* spp

Morus belongs to the family *Moraceae* and comprises about 10-16 species of deciduous trees, commonly known as mulberry and locally known as tuth (in Arabic) and tul (in Kashmiri). The most common species of mulberry in India include *Morus alba, Morus indica* and *Morus nigra* (shah tul). Many varieties/ clones have also been developed worldwide, like Gosherami, Chinese white, Ichinose, etc. SKUAST-K has also developed SKM lines of mulberry, showing great promise. In the Kashmir valley, mulberry is mostly grown for silkworm rearing and it also provides fuel wood, fodder and fruits.

Propagation

Mulberry is mostly raised from cuttings in nursery condition as follows:

1. Disease- free cuttings of 15 -20 cm length and 1.2 -1.5 cm diameter with 3-4 active buds are to be selected from 8-10-month-old shoots.
2. The cuttings should be treated with 0.02% Bavistin solution at the cut ends for half an hour to ensure protection against fungal diseases.
3. Well- punctured polythene bags (4.5 inches diameter and 11 inches height) should be filled with rooting medium comprising sand, soil and well- decomposed FYM in the ratio of 6:3:1. The treated cuttings should be gently inserted in the polybags without damaging the bud, keeping the uppermost bud exposed and finally, these bags should be placed in the polyhouse.
4. The insertion/plantation of cuttings should be done between the last week of March and the first week of April.
5. Optimum hygrothermic conditions, viz., 25-30°C temperature and 75- 80% humidity should be maintained in the polyhouse.
6. Irrigation should be carried out, as and when needed. However, there should be proper drainage of water from the polybags, as otherwise, the saplings would decay.
7. Fertigation should be carried out after 40 days.

8. After 75 to 90 days, the saplings should be transplanted to the main field They should be planted at a distance of at least 9″ x 9″ from each other.
9. Immediately after transplantation, sufficient irrigation should be provided to enable the saplings to get established. During hot days, the frequency of irrigation should be maintained as per requirement. In addition, the field/ nursery should be kept free of weeds as far as possible.
10. 1-2 year-old saplings are fit for transplanting in the field.

Management

Mulberry is prone to the attack of pests like *Glyphodes pyloalis* that can cause defoliation on a large scale. Initially, when the infestation is less, the infested leaves can be plucked and collected for burning. In case the infestation is very high, any type of contact insecticide can be prescribed.

2. *Ulmus villosa*

Ulmus villosa belongs to the family *Ulmaceae* and is commonly distributed in the western Himalayas and endemic to the valley of Kashmir at an elevation of 1200-2500 m. It is commonly known as Cherry-bark Elm or Marn Elm and locally known as Bren (in Kashmiri). The tree comes into flowering early in spring. Flowers are borne on the leafless twigs in spring. They are minute and reddish in colour and the fruit is winged, rounded and peppery, 9-13 mm in diameter, with a seed in the centre. (most of the seeds were unfilled.?) Studies have been conducted to find out the optimum time when a large number of viable and germinated seeds can be collected. They concluded that the 3rd to 4th week of March is the most suitable time for the collection of *Ulmus villosa* seeds in the valley, depending upon environmental conditions, particularly temperature. Therefore, seeds should be collected at the proper time. Seeds collected at maturity showed a higher weight of 10.14 g per thousand seeds (Bhat et.al., 2007).The seeds of *Ulmus villosa* do not have any kind of dormancy. Therefore, they do not require any kind of treatment. Bhardwaj and Mishra (2005) studied that in *Ulmus villosa*, treated cuttings with chemical formulations of 0.4% IBA + 0.2% p-HBA + 5% sucrose + 5% captan showed maximised rooting (82.0%), survival (80%) and primary root number (14.7). Their study also revealed that rooting success and root number were better in propagules set in February, rather than in July. Cuttings from seedlings rooted better with more roots, ensuring better survival, than cuttings from mature trees.

Propagation

1. The softwood and hardwood cuttings should be taken from phenotypically superior trees and should be treated with 200 ppm IBA for 24 hours.

2. After that, they should be planted in well- prepared beds under controlled conditions.
3. In one growing season, the seedlings attain a height of nearly one metre.
4. The seedlings can be uprooted from beds and planted at the plantation site after one growing season, if the area of plantation is properly fenced.
5. Otherwise, thinning should be done and the seedlings can be transplanted in transplant beds at a spacing of 30cm x 30cm for one more season.
6. The tree species grows under water stress conditions, but displays vigorous growth under an assured supply of irrigation.

3. *Quercus robur L*

Quercus robur L., commonly known as English oak, belongs to the family *Fagaceae.* English oak is a large deciduous tree with a broad and spreading crown, a short, sturdy trunk and deeply fissured grey- brown bark. In a nursery study, seedlings of English oak were raised in open beds of size 1 x1m, root trainer (150cc, 250cc and 300cc) and polybags (150cc, 300cc and 7"x5"), in a medium of soil, sand and FYM (2:1:1). The seedlings raised in nursery beds performed better than the containerised seedlings (Iqbal, 2012). A couple of experiments were conducted for the propagation of *Quercus rober L.* by stem cuttings in the Kashmir valley. In the first experiment, hardwood cuttings were treated with Indolebutyric acid (IBA), having concentrations of 5,000, 10,000, 15,000 and 20,000 ppm in talc and Naphthalene acetic acid (NAA), having concentrations of 500, 1000 and 1500 ppm in talc. In the second experiment, softwood cuttings were taken and treated with Indolebutyric acid (IBA), having concentrations of 5,000, 10,000, 15,000 and 20,000 ppm, quick- dipped for 5 seconds and Naphthalene acetic acid (NAA), having concentrations of 500, 1000 and 1500 ppm, dipped for 24 hours and in both (non- treated) as control and placed under mist conditions. It was found that English Oak can be propagated through cuttings and auxin treatment is imperative. Indolebutyric acid (IBA) with a concentration of 10,000 ppm showed the best results with the highest recorded rooting of 51.30% in the case of softwood cuttings. For hardwood cuttings too, Indolebutyric acid (IBA) with a concentration of 10,000 ppm showed good results with the highest recorded rooting of 29.70%. However, control and NAA treatments in both softwood and hardwood cuttings could not induce rooting at all (Iqbal *et al.,* 2014).

References

Antony Joseph Raj and Lal S.B. 2014. Agroforestry Theory and practice. Scientific Publishers (India) 929pp.

Batool, N., Bibi, Y. and Ilyas, N. 2014. Current Status of Ulmus wallichiana: Himalayan endangered Elm. Pure Appl. Bio, 3(2): 60-65.

Bhardwaj D. R. and Mishra, V. K. 2005. Vegetative propagation of Ulmus villosa: effects of plant growth regulators, collection time, type of donor and position of shoot on adventitious root formation in stem cuttings, New Forests, 29 (2): 105–116.

Bhat Gh. Mohi-ud-Din., Khan M.A and Mughal A.H (2007). Seed maturity indices in Elm (Ulmus villosa). A multipurpose tree species in Kashmir valley. International Journal of Environment, Ecology and Conservation 13 (3): 2007, 473-476.

Burley J and von Carlowitz P 1984 Multipurpose Tree Germplasm, ICRAF, Nairobi, Kenya,298p.

Chundawat B S and Gautam S K 1993 A textbook of Agroforestry. Oxford & IBH Publishing Co. Pvt. Ltd. New Delhi.

Forest/Fuelwood Research and Development Project 1992. Growing multipurpose trees on small farms, Bangkok, Thailand: Winrock International, 195 pp

Iqbal, J., Dutt,V, Ahmad, H., Bhat G.M. and.Khan P.A. 2014 Propagation of Quercus rober L. (English oak) by stem cuttings in western Himalayas (Kashmir). Environment Conservation Journal. 15(1&2)185-189.

Iqbal, J. 2012 Status and Propagation of Oak (Quercus spp.) in Kashmir valley. M.Sc. thesis, SKUAST- Kashmir, J&K.

Masoodi, T.H., Masoodi., N.A., Bhat,G, M, Dutt, V. 2013. Ecological and Economic Security from Willow based Land use Systems in Kashmir, Faculty of Forestry, SKUAST-K, Sopore,Baramulla.pp.123

Melville, R. and Heybroek, H. M. 1971. The Elms of Himalayas. Kew Bulletin 26 (1): 5-28. Mughal A H and Mugloo J.A. 2015 Important trees of Kashmir Valley. SKUAST –Kashmir, J&K.

Mughal,A.H. 2017. Agroforestry and scope of mulberry based system in Kashmir Valley. Training Manual on "Integrated module for mulberry based farming systems" for ten days training programme 19th -28th Dec., 2017, TSRI, Mirgund, pp. 33

12

Nursery Raising Techniques of Wild Apricot (*Prunus armeniaca*) Under Temperate Conditions of Kashmir Valley

A.H. Mughal, J.A. Mughloo, P.A. Khan, Rameez Raja and Mehraj ud Din Dar

Faculty of Forestry, SKUAST-K , Benhama, Ganderbal-191201, Jammu and Kashmir, India

---

Apricot is an important fruit grown in the dry temperate and mild-hill regions of India. The fruit belongs to the family *Rosaceae* and genus *Prunus*. The original apricots are native to China, whereas wild apricot, which is known as 'Zardalu', is indigenous to India. Wild apricots are found at a low altitude of 1500 m and as high as 4000 m a s l. The cold hardy trees are hardier than peach and almond and can tolerate winter temperatures as low as -30°C. They are drought- tolerant and can be grown in a variety of soils, but deep, fertile and well- drained loamy soil with a pH value of 6.0 to 6.8 is suitable for their growth and development. Apricots are found the world over at low altitudes of 150 m and as high as 4000 m a s l. The species and varieties of apricot range in areas of adoption from the cold winters of Siberia to the cold, arid conditions of Japan and Eastern China. It is also found in Turkey, Spain, Italy, USA, France, Greece, Syria, Morocco and India.

The distribution of cultivated apricot, its wild forms and allied species is reported in the temperate zone of Asia between 33^0 and 75^0 E longitude and 530 and 300 N latitude. In the state of Jammu and Kashmir, wild apricot is grown throughout the state, except in Jammu district and major apricot- growing belts are located in Leh and Kargil districts, where it covers a total area of 839 and 1873 ha respectively. Wild apricot is commercially propagated by grafting or budding and T-budding. Tongue budding is used for multiplication. The selected superior plants of wild apricot can be grafted on to the root stock of the Chuli. For raising the rootstock, seeds are collected from fully ripened wild apricots.

Different Steps Involved in Nursery Raising Techniques

Seed collection

Seeds should always be collected from superior trees in which fruit and kernel size is big. Small- sized fruits and seeds should be avoided. Seeds of ripe fruits should only be collected for the purpose of raising seedlings. Unripe and fallen fruits should be avoided. After collection, the seeds should be thoroughly washed as any amount of pulp left on the stones may often cause a loss in the viability of the seeds. The pulp can be removed simply by rubbing or washing. Thereafter, the seeds should be cleaned before they are dried in the sun. Seeds, if stored without proper drying, are prone to damage by insects and pests. Sun-dried seeds should be stored in gunny bags or seed containers and stored in a cool dry place till the time of sowing in the nursery. On an average, 850-900 seeds weigh about 1 kg.

Seed Testing

In order to discard empty seeds right at the time of sowing, the seeds collected are subjected to a float test. For this purpose, the seeds are kept in a bucket of water for at least 48 hours. Heavy seeds having healthy kernels sink to the bottom, while empty seeds keep floating even after 48 hours and should be removed and discarded.

Pre Sowing Treatment

Seeds need stratification for a duration of 45 days at 40 C to break dormancy. The stratification medium should be moist throughout the period and should have proper aeration. Generally, the seeds are kept underground in layers of sand during winter. Seeds that are sown in autumn do not require stratification as they are stratified in the seed beds under natural conditions during the winter and they start germinating in the spring. After stratification, the seeds need to be soaked in @ 5.0 ppm Kinetin or 500 ppm GA3 solution for a period of 24 hours before sowing, for speeding up the germination. The three growth regulators at different levels, viz., GA @ 250,500,750 ppm, thiourea @ 0.2, 0.4 and 0.6% and KNO 0.3 and 0.4 % have been found to be effective and increase the speed of germination up to 80%.

Seed Sowing

Seed sowing is done in lines after making furrows of about 25 cm. A distance of 25 cm is maintained between the furrow lines. Seeds are, thereafter, placed in the furrows at a distance of 15 cm. from each other. After placing the seeds, the furrows are closed. The seed beds are then mulched with 5-6 cm thick grass and light irrigation is applied. Germination starts in the spring in the month of March. Regular weeding, hoeing and irrigation is carried out throughout the growing season and the seedlings attain a height of 4-5 ft in one growing season.

Land Preparation, Spacing and Planting in Apricot Farming

Dig the pits just 1 month before planting. Each pit should be 1m× 1m×1m and filled with a mixture of soil and 60 kg of well-rotted farmyard manure (FMY), 1 kg of single super phosphate and 10 ml of chorapyripos solution (10ml/10 litres of water). As usual, on slope lands, the contour planting system and on flat lands, the square or triangular planting method should be followed. The plant spacing depends on the variety and soil. The apricot plants are generally spaced at 6m × 6m. Transplant the nursery- raised 1- year-old seedlings to the main field and place them in the centre of the pit. To conserve the soil moisture and check the weeds at basins, mulch the plant basins with 10 to 15 cm thick hay. Make sure to irrigate the plants immediately after placing them in the pits of the main field.

Manures and Fertilizers in Apricot Farming

In apricot farming, both organic and chemical fertilisers are required for tree growth and good yield. The fertiliser requirement depends on the soil type, tree age and cultural practices.

Under irrigated conditions, for each matured apricot tree (7 years and above), a mixture of 40 to 45 kg well-rotted farmyard manure, 500 grams of N, 250 grams of $P_2 O_5$ and 200 grams K is preferred. The FYM should be applied along with a full dose of P and K in the month of Dec. to Jan. As far as nitrogen application is concerned, it should be applied in two split doses - half the dose should be applied 3 weeks before flowering and the remaining after a period of 4 weeks. Under rainfed conditions, the second half dose of N should be applied at the onset of the monsoon.

Weed Control

In apricot farming, the pre-emergence application of atrazine or Diuron@4.0kg/ha and post- emergence application of Glyphosate @ 800 ml/ha or Gramaxone @ 2 litres/ha is more effective to control the weeds. Mulching also checks the weed growth.

Insect and Pest Management

White Grubs

They attack roots and cause wilting, thereby resulting in the death of seedlings in the nursery. Forate 10g or aldicarb 10g should be mixed in the soil for the control of grubs.

Aphid

Nymphs and adults suck the sap from the leaves and petioles, causing leaf curl and distortion.

Hairy Caterpillar

This feeds on leaves, resulting in defoliation, if the infestation is heavy.

Aphid and hairy caterpillars can be controlled by spraying chlorpyriphos @ 0.02% or endosulfan @ 0.5% or Dimethoate 0.30%..

References

Cai-ru, GUO, Zhen-long, WANG and Ji-qi, LU.(2010). Seed germination and seedling development of Prunus armeniaca under different burial depths in soil. Journal of Forest Research 2010. 21(4):492-496.

Khan, M.A, Mughal, A.H. Mugloo, J. A,Sofi, P.A. and Mir, A.A.(2009). Seed source variation and correlation studies in wild apricot (Prunus armeniaca) genotypes of Kashmir and Ladakh region. Environment & Ecology 27(1A): 472-476.

Kumar, A and Shahnaz, E. (2013). Effect of growth regulating substances on stratification of wild apricots (Prunus armeniaca) kernels under Kashmir conditions. Progressive agriculture.2013. Vol: 13, Issue:2.183-187.

13

Important Broad-leaved Agroforestry Tree Species of Kashmir Valley

G.M. Bhat, A.R. Malik and Nazir A. Pala

Faculty of Forestry, Benhama, Ganderbal
SKUAST-Kashmir 191121, Jammu and Kashmir, India

Introduction

Agroforestry denotes a sustainable land and crop management system that strives to increase yields on a continuing basis, by combining the production of woody forestry crops (including fruit and other tree crops) with arable or field crops and/or animals, simultaneously or sequentially, on the same unit of land and applying management practices that are compatible with the cultural practices of the local population. In Jammu and Kashmir, agroforestry systems have been practiced since centuries and passed from generation to generation. These systems were built on the foundation of protecting and planting trees. These systems in the past have made the hill people self- sufficient and well-nourished. It was well said by Leaky (2001) that agroforestry is now being seen as an alternative paradigm for rural development worldwide that is centred on species-rich, low input agricultural techniques including a diverse array of new indigenous tree crops, rather than on high input monocultures with only a small set of staple food crops. However, there is need for improvement in these systems so that they will remain sustainable and more and more adaptable for the rural people. The state of Jammu and Kashmir has a total geographical area of 1,01,387 km2, comprising 22 districts. The Kashmir region of Jammu and Kashmir represents a typical temperate ecosystem. The physiography of Kashmir is a fertile basin of valley measuring 187 km to 115 km along the Srinagar latitude. Mughal and Khan (2007) reported that there were only seven systems prevalent in Kashmir province. Bhat *et al.*, (2010) reported that seven common traditional agroforestry systems are practiced in Kupwara district of the Kashmir valley. There are different important multipurpose tree species which are grown by farmers in different agroforestry systems. Some of them are described below.

Poplar- based agroforestry systems: Poplars are an important multipurpose, fast-growing broad-leaved tree species of the Kashmir valley. Different cultivars of poplars are found growing in the Kashmir valley under different agroforestry systems. They are mostly grown by the farmers on the boundaries of their lands, as pure stands, on the banks of streams and nallas and river banks. They are grown for their use in light construction, ply board industries, packing cases for fruits and as fodder for cattle. They are also used as windbreaks and shelterbelts and their twigs are used as fuel wood. The Poplars are the most suitable tree species for the Kashmir region. Nearly 70-80 % of the demand for the fruit boxes is met by this species and now, even the wood is used for construction purposes. The farmers of the valley grow various species of poplars viz., *Populus deltoides, P. nigra, P. balsemifera, P. alba* and *P. ciliata. P. deltoides* is the most common which is recommended to be planted in two ways:

i) Along the boundary/bunds of the field

ii) In pure stand of Poplars

The spacing of trees is maintained depending upon the objective of raising plantation. For commercial purposes, i.e., for packing cases, a spacing of 2 x 2 m is to be maintained, while for agroforestry systems, 4 x 4 m and above is recommended. Also, for fencing and shelter belts, closer spacing than the above-mentioned is recommended.

Nursery Raising of Poplars

- Nursery sites for poplars should be sandy loam, free from boulders and stones
- Good irrigation facility should be available at nursery sites
- Nursery sites should be well- drained and waterlogged areas should be avoided
- A soil P^H of 5.5 to 7.0 is the most suitable for raising popular nursery

Preparation of Nursery Beds

- The soil should be ploughed up to a depth of 50 cm and thoroughly pulverised.
- Organic manure is to be applied @ 20 tones ha.
- Phosphorus is to be applied as a basal dose @ 120 kg ha.
- The soil should be thoroughly worked after adding organic manure and phosphorus.

- After ploughing and mixing of fertilisers and manure, the land should be levelled and beds laid out in the east and west directions.
- The size of the beds should be 5 m x 1.5 m.

Preparation of Cuttings and Their Planting

- One-year-old branch cuttings should be selected from phenotypically superior and disease- free trees.
- The size of the cutting should be 22cm with at least 4 buds and should be from the semi- hard portion of the branch, with a diameter of 1.5 cm and above.
- The cuttings should be prepared after leaf fall in the month of December after giving a slanting cut on both sides.
- The prepared cuttings should be buried under the soil till nursery planting , starting in the last week of February to March.
- The cuttings should be inserted in the nursery beds at an angle of 45° at a distance of 25 x 25 cm. 16 cuttings can be accommodated in a bed size of 1 x 1m.
- After inserting the cuttings in the bed, the soil around them should be compacted and watered

Management of Nursery

- The nursery beds should be regularly watered, preferably through flood Irrigation.
- Urea should be applied in two split doses of 75kg ha. in the month of April and the last week of June.
- To ensure good and vigorous growth of the seedlings, weeding of nursery beds should be carried out regularly, without disturbing the cuttings.
- Singling of shoots should be carried out in the month of July-August and vigorously growing shoots be retained.
- During one growing season, the seedlings on the best site attain a height suitable enough to be transplanted in the field.
- Interested farmers can raise saplings in a polyhouse too

Disease/Pest Management

A number of diseases and insect pests, such as leaf spots, blight/defoliators etc. attack seedlings during the developmental stage in the nursery. They can be easily managed by using fungicides and pesticides. 0.1% drenching with Metalaxyl or carbendazim followed by foliar spray with mancozeb@0.3 % will take care of disease, whereas insects can be controlled by spraying endosulphan @ 0.1% or chloropyriphos @ 0.1%.

Willows (Salix species)

Willows, locally called Veer, have been growing in the Kashmir Valley since time immemorial. They are found growing as pure stands, on canals, wastelands, irrigation channels, roadsides, boundaries of fields and bunds and also scattered under different agroforestry systems. The large-scale and commercial plantation of this species around the Wular Lake was undertaken by the forest department in 1917. Thereafter, other wetlands located at Harran, Hygam, Hokersar, Mirgund, Mamandangi, Gund Jehangir and Shahgund were brought under the willow plantation and at present, about 1,400 km^2 of land is under its cultivation. These plantations were primarily meant to supply the fuel wood requirements of the towns of Kashmir.

Utilization of Commercial Willowsin Kashmir

- Cricket bats
- Packing Cases
- Basketry and Furniture
- Biomass Energy
- Soil Conservation_
- Phytoremediation
- Production of Fodder_
- Medicinal properties and rooting Stimulant

Propagation (Nursery raising)

Cricket bat willow is propagated by using sets – carefully selected cuttings (from quality stock) - which are pushed into the ground in holes prepared with a metal bar. These sets are planted from December to February with a crowbar or similar tool. A 45- degree angle is made by cutting a portion of the basal part of the set and it is pushed to the bottom of the hole, in which, a little water has been poured.. It is essential that the sets are in the upright position. 'Firming up' should take place in March, taking care not to break the newly- formed roots. After one year's growth, the rooted cuttings are lifted from the ground, cut back to a single stout bud and then replanted in the nursery. The solitary bud grows as a straight 'set' and the 'rooted sets' are then planted in the field at a wide spacing, usually > 4x6 m apart.

Direct Plantation of Poles/Mawas

Planting poles or mawas is another method of raising cricket bat willow trees in Kashmir. Compared to cuttings (40-50 cm long and 15-25 mm in diameter) which are usually raised as nursery plantings, the poles/mawas are normally 3 m long with an approximate diameter of 30 mm at the top end and 55 mm across the base and planted directly at the plantation sites.

Planting willow Poles/Mawas

Cricket bat willow plantation

Essentials of Pole Planting

- It is crucial to select material that is freshly- cut, straight, well- grown with a suitable diameter and without knots. Any deviation will result in less vigorous growth, besides taking much time to grow into useful trees.
- Be sure to label the planting material with the correct clone name. Keep a plan of what clone has been planted at each particular site, for future reference.
- An essential precaution to be taken when transporting poles is to protect the bark from damage. Use straps, ropes and protective pads, rather than chains, when securing a load of poles and unload them carefully.
- The best season for planting willow poles/mawas is from mid- Feb. to March, when soil moisture levels are high, temperatures are lower and planting material is dormant. It is important to plant poles into a moist

soil. It is also vital to compact the soil after planting to ensure that the pole is firm in the ground and in good contact with soil moisture.

- A three-metre pole will need a planting depth of 80 cm (up to a third of pole length). Thus, depending on the condition, an average planter can plant 50 to 80 poles a day. It is always better to take as many mawas out to the planting site as you can plant in one day so as to protect them from desiccation.
- Wherever possible, a post hole borer, with an auger just big enough to create the right-sized hole for a pole, should be used. Although it involves carrying another tool around, most contractors use a separate crowbar or a motorised auger. In almost all cases, a small pilot hole with a diameter less than that of the pole is formed, into which, the pole is driven to the required depth. This gives excellent contact between the pole and the soil.
- Before planting, slice the lower end of each pole at an angle of 40-60 degrees so that it can absorb water and then soak these ends for 8-12 days, preferably using fresh-flowing water, to ensure they are full of water at planting time.
- Ensure that the poles are planted in suitable sites, preferably in moist soils or depressions where they will grow well and also that the poles are pointed the right way up.
- If the soil is loose around the mawas, then re-ram them tightly. Re-ramming should be done carefully using a regular fencing rammer and care should be taken to avoid damaging the bark and emerging roots.

Pole Protection During Establishment

Sleeves are recommended to prevent grazing stock and wild animals from eating the bark and killing the tree. In practice, cattle should not be grazed in blocks with newly planted poles for at least one and preferably two growing seasons. Two commonly used protective sleeves are commercially available for use on poles – 'Dynex' and 'Netlon' sleeves. These are designed to prevent grazing stock from eating the bark off the poles. 'Dynex' sleeves are made from an extruded plastic tube that fits over each pole. The heavy plastic sleeve helps to protect the poles from stock damage during the establishment years and can minimise moisture loss when dry winds prevail. The new 'Dynex' type has a diamond cut zipper, which splits as the tree's diameter expands. 'Netlon' sleeves are made from plastic mesh. They are suitable where sheep are the only stock having access to the poles. Place the sleeve on a pole before planting, pulling it up over the pole base to around 70 cm from the butt end and secure it there, using two vertically aligned staples placed about 150 mm from either end. In this way, a 'Netlon' sleeve will act as a planting depth guide. As the pole grows into a vigorous tree, the sleeve will break down and fall away.

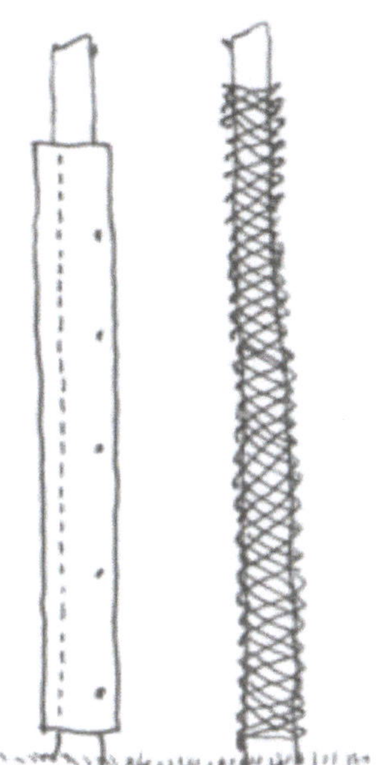

Staking

Trees planted in windy or exposed locations need to be staked to prevent them from falling or breaking. For most trees, a low stake is preferred so as to let the tree move naturally. However, if winds of high speed are more prevalent, use a high stake. For trees more than 12 feet tall, use two low stakes on opposite sides of the tree. The ties used need to accommodate growth and not cause bark damage due to friction. Buckle-and-spacer ties, which are expandable and have a protective spacer, can be used. Ties without spacers should be formed into a figure eight to create padding. While staking a tree, provide enough leeway so that the tree can move back and forth in the wind to develop stronger roots. If the tree cannot move back and forth, its roots will not develop and the tree might fall during a storm.

Tending

Growers have to tend the trees carefully if they are to sell them to the manufacturers of cricket bats. The most important tending operation right from the early stages of growth is the removal of small epicormic shoots on the stem. If allowed to grow, these shoots form a pin-sized knot in the wood which weakens it and hence, reduces its value. Trees that have been neglected for one year only during their lifetime have little or no commercial value. Thus, the side shoots should be rubbed off up to the height of at least 84 inches (213.5cm) from ground level. The first five years are the most difficult to carry out maintenance, after which, the bark will harden up and the side shoots become less evident.

Harvesting of Trees and Conversion into Clefts for Crickets Bats

Cricket bat willow trees are normally purchased standing, with the felling, extraction and clearing done by the contractor/ unit holder. This species is fast growing and it is possible to obtain trees of suitable diameter with a 7-10 ft clear bole in 18 years from the time of planting the sets. Thus, the trees are normally felled between 15 and 20 years of age or once they attain at least 58 inches (147.5 cm) girth at 4' 8 inches (142.3 cm) height from ground level. The life cycle of a commercially grown tree is ideally 15 to 20 years. They can be allowed to grow beyond this stage provided they remain healthy and can grow to a size well in excess, to obtain more clefts.

Both the tree species, Salix and Poplar, are recommended for growing in the Kashmir valley under different agroforestry systems which are tabulated below:

S.No	Agroforestry system	Functional Unit Kharif	 Rabi	Major output/ Function
1	Agrisilviculture	Maize, French Beans	Oats and Mustard	Food, fodder, Raw material for plywood, sports industries
2	Agri hortisilviculture	Beans, Moong, Vegetables. Fruit trees-Apple, Cherry, Almond, Plum, Apricot, (perennials)	Fodders-oats, orchard grass, Tall fescue	Food, Fruits, Fodder, Small timber, Raw materials for industries
3	Home gardens	Agriculture-vegetables, maize, Horticulture- Pear, peach, Apple, Plum, Fodder (M.P.Cherry, Trifolium, Animals and Birds Cow, goat, Sheep, ox, Buffalo, Poultry?)	Vegetables-Carrot, Spinach, Kail, Fodder- Oats	Fodder, Fruit, Milk, Vegetables, Meat, Eggs, Skin,
4	Boundary Plantation	Paddy, Maize.	Mustard, Fodder Crops	Live fence, fodder, poles, Timber, Raw material for industries windbreaks

In addition to this, many tree species are grown mixed with these two important multipurpose species. They include Ailanthus, Horse chestnut, Black locust, Acer, etc.

Elm-based Agroforestry Systems (Alley Cropping for kareva lands of Kashmir)

Elm is commonly distributed in the western Himalayas, from northern India to Nepal, at an altitude of 900-3000 m .The species is scattered from central Nuristan, along the Himalayas, through Kashmir and into Nepal . In Kashmir, Elm is grown for its multifarious uses and has become a preferred tree species in our agroforestry system. It is often planted around villages, along the banks of streams, on dry ridges and sloppy lands etc. It is also grown around the boundaries of fields in some places. Elm grows under moisture stress conditions of the Kandi area in the Kashmir valley. Its timber is used in light construction, planking cases, furniture and handles for agriculture tools. Its leaves are valued for good fodder and bark, for sandals and ropes. The bark also contains 0.76 % tannin. Its branches are used for firewood..Elm bark is used in a recipe for an ointment to heal broken bones and has also been reported to be used for the treatment of cuts in Mussoorie. The tree comes into flowering early in the spring season. Flowers are borne on leafless twigs in the spring. They are minute, reddish in colour and the fruit is winged, rounded and peppery, 9-13 mm in diameter, with a seed in the centre. (most of the seeds were unfilled .?) Studies conducted to find out the optimum time when a large number of viable and germinated seeds can be collected revealed that the 3rd to 4th week of March is the most suitable time for the collection of *Ulmus wallichina* seeds in the valley, depending upon the environmental conditions, particularly temperature. Therefore, the seeds should be collected at the proper time.

The seeds of *Ulmus wallicina* do not have any kind of dormancy. Therefore, they do not require any kind of treatment. Seeds collected at maturity showed a higher weight of 10.14 g per thousand seeds. The tree has been recommended for plantation in the karewa lands with 25 to 30° slope of Kashmir under alley cropping. Different crops can be grown in the alleys as kharif crops like tomato and potato, followed by rabi crops like garlic.

Nursery Raising of Elm

- Elm is a spring, seeded species and the 3rd to 4th week of March is the most suitable time for the collection of Ulmus wallichina seeds in the valley, depending upon the environmental conditions, particularly temperature.

- Nursery sites for Elm should be sandy loam, free from boulders and stones.
- Good irrigation facilities should be available at the nursery sites.
- Nursery sites should be well- drained and waterlogged areas should be avoided.
- Soil pH of 5.5 to 7.0 is most suitable for nursery raising.

Preparation of Nursery Beds

- The soil should be ploughed up to a depth of 50 cm and thoroughly pulverised to obtain good tilth.
- Organic manure should be applied @ 20 tons/ ha.
- The soil should be thoroughly worked after adding organic manure.
- After ploughing and mixing of fertilisers and manure, the land should be levelled and the beds laid in the east and west directions.
- The size of the beds should be 1 m x 1m.
- The fruit is called samara, which is reddish- brown in colour, having a seed in the centre. Seeds can be collected from phenotypic ally superior trees and should be sown immediately after collection.
- Pre- sowing irrigation should be given to the nursery beds.
- The seeds should be broadcasted as they are very small.
- The seeds do not have any kind of dormancy and start germinating within three days of sowing.
- The nursery beds should be irrigated as per need with the help of fountains and buckets.
- Since the seeds are small, they should not be sown deep.
- Seed germination percentage ranges between 85 to 90%.

Vegetative Propagation

- Elm is a hard to root species. Studies have shown that its vegetative propagation by means of cuttings that have been treated with plant growth regulators enhances rooting percentage.
- Among different plant growth regulators, IBA @ 200 ppm enhanced rooting percentage to 28% as against 3% in control.
- The hardwood cuttings should be taken from phenotypically superior trees and treated with 200 ppm IBA for 24 hours.
- After that, they should be planted in well- prepared beds under controlled conditions.

- In one growing season, the seedlings attain a height of nearly one metre.
- The seedlings can be uprooted from the beds and can be planted at the plantation site after one growing season, if the area of plantation is properly fenced.
- Otherwise, thinning should be done and the seedlings can be transplanted in transplant beds at a spacing of 30 x 30 cm for one more season.
- The tree species grows under water stress conditions, but has vigorous growth under an assured supply of irrigation.

Under block plantation, a spacing of 3 x 3 m is recommended and 1,111 plants can be accommodated in one ha. of land. The spacing can be increased or decreased from the above- mentioned spacing, depending upon the objective of planting.

Fruit Tree Based Agroforestry Systems

The main focus of such systems is fruit production. Different kinds of fruit trees are grown in the valley, viz., apple, almond, peach, pear. apricot, cherry etc. but among them, apple trees are grown in all districts for both commercial and domestic purposes. It was observed that the areas suitable for agriculture (paddy) are converted into orchards. Horticulture is considered as the backbone of the economy of the state. Traditionally, farmers raise maize as a kharif crop and wheat as a rabi crop. Since both crops were heavy feeders, it was the need of the hour to develop a model which was high in productivity as well as sustainable and adaptable for the farmers. Therefore, a system was developed by the scientists of SKUAST, in which, different fodders such as white clover, Lucerne, orchard grass, medicinal plants like *Artemesia* and agriculture crops like beans and peas were sown. The system has shown promising results. The highest fruit yield was recorded in apple + white clover (24.2 kg/tree or 14.76 t/ha and 6.41 kg/tree) followed by apple + beans-pea (22.16 kg/tree or 13.51 t/ha.). The control (apple + natural sward) yielded the lowest fruit yield (3.5 kg/tree or 2.13 t/ha). Also, maximum biomass (10.31 t/ha) and carbon stock (4.63 t C/ha) was removed from the treatment. Apple+Lucerne was closely followed by apple + beans-pea (9.77 t/ha and 4.39 t C/ha). The control (apple + natural sward) recorded the minimum biomass (5.07 t/ha) and carbon stock (2.28 t C/ha) that was removed. Consequently, the apple +Lucerne registered the highest amount of carbon sequestered, i.e., 9.56 t C/ha and Co equivalent of 35.08 t/ha. Our study revealed that treatment (apple +Lucerne) would conserve 1.02 million tons of carbon and 3.76 million tons of CO_2 equivalent. This model is adopted in every part of the valley by progressive farmers and has increased their income and improved the physical properties of the soil. It has also resulted in a source of carbon sink.

References

Gh. Mohi-din Bhat, Mugloo, J.A, T.A Rather,Amarjeet and A.A. Mir (2010) Traditional Agroforestry system plasticized in Kupwara; A border and backward district of Jammu & Kashmir. Indian Journal of Forestry 33(2);173-176

Leaky, R.R.B.2001 Agroforestry for Biodiversity in farming Systems In: Collins W.W and Quaiset C O eds. Biodiversity in agro ecosystems. New York, CRC press. Pp. 127-145

Mughal, A. H & Khan, M .A. (2007). An over view of agroforestry in Kashmir valley. In: Agroforestry systems and practices. P 43-54 Eds. Sunil Puri & Pankaj Panwar. New India Publishing Agency New Delhi

14

Technological Advances in Plantation Techniques of *Populus deltoides Bartr.* for Problematic Sites in Temperate Regions

Tahir Mushtaq

Technical Officer, RCFC North, FoA, Wadura, Jammu and Kashmir, India

Introduction

Land is one of the most important resources on which human beings depend. The rate of soil degradation is continuously increasing with the advancement of science and technology, industrial expansion, urbanisation and population explosion. The most important cause of land degradation is the destruction of forests and other vegetation in sloping lands, riverbanks and other areas sensitive to damage (Pramod and Mohapatra, 2012). Anthropogenic activities like over grazing, wood cutting and burning have intensified land degradation, resulting in soil deterioration all over the world (Jin, 2002). Vegetation acts as a protective cover against the forces of wind and water, protecting the soil from being washed or blown away and preserving the physical and hydrographic balance of nature (Jain and Singh, 1998). Plantations can conserve soil on degraded lands by reducing nutrient loss, increasing soil organic matter and improving the soil texture (Thapa, 2003). The plantation forestry is an ecologically as well as economically more viable option than traditional forestry. Plantation technology of fast growing species, mainly Poplar and Eucalyptus, has been taken up by farmers and institutions to boost wood production in the world. Plantation forestry has the potential to augment a farmer's income substantially. Tree plantations present an economically attractive alternative to natural forests and may also be a practical option for bringing the degraded lands to the production option. Plantation forestry has been shown to contribute considerably in terms of carbon sequestration, increased soil organic carbon and the conservation of biodiversity (Bremer and Farley, 2010).

In India, a major portion of the land is affected by different types of soil degradation (soil and water erosion, soil salinity, alkalinity, acidity, etc.). Large areas in the country have been rendered useless as a result of soil deterioration and the problem of low productivity due to soil losses is much larger as one can think. The problematic soils cover an area of 24 million hectares and accounts for 7.30% of the total geographical area of the country. It presents a serious threat to achieve the target of higher crop productivity. The reclamation of the problematic lands of India is becoming important because of the increased demand for food, fodder, fuel and shelter (Banerjee and Raza, 1992). The National Wasteland Development Board (NWDP) was set up with the objectives of protecting the good lands from further degradation, increasing tree and other green cover on the problematic lands, promoting fuelwood and fodder trees and involving people in afforestation programmes. The most important way to reclaim problematic soils is to plant trees (Tomar *et al.*,2003). There is growing concern in India about the degradation in the quality of forests and the significant reduction of the forest tree cover.

Jammu and Kashmir is predominantly a hilly state with a forest cover of 20,230 km^2 and accounts for 19.95% of its total geographical area of 1,01,387 km^2. About $^{2/3rd}$ of the state's total area is under recorded forest and a substantial part of it is non-conducive for growth since it is under permanent snow, glaciers and cold deserts (Digest of Forest Statistics, 2012). The total problematic area of the state is about 45.70%. This area is prone to soil erosion and other forms of degradation, leaving behind the denuded and degraded soils with poor service for mankind. The degradation of forest resources in Jammu and Kashmir has been accelerating, owing to the rapid growth of population coupled with the development of agriculture and urban construction. The situation is further aggravated by the faulty forest management practices in vogue. As a result of forest degradation, environmental problems including soil erosion and loss of biodiversity are being experienced and natural hazards are occurring with increasing frequency. Now, the government of Jammu and Kashmir is focusing on the conservation of forests and restoration of problematic areas existing in the state.

To maintain the fertility of degraded soils, it is essential that the lost nutrients are made equally good by artificial means. Fertilisers play a vital role in boosting the initial growth and development of plants (Moscatelli *et al.*, 2008). However, the optimum requirement for different fertilisers varies with the species as well as the prevailing soil fertility status. In the developed countries, fertilisation is the conventional tool used by foresters for maintaining higher productivity, whereas, the application of chemical fertilisers in plantations is not very common in developing countries like India. It is well- documented that too much or too little N, P and K may result in poor establishment and growth of the stands. Optimum fertilisation extended the fast growing period of the tree as well as the survival percentage of the outplanting (Jagger and Pender, 2003). Among the various nutrients, nitrogen is the major element required for the growth of all plant species and is always available in limited quantities in the soil. Nitrogen, being a major plant growth nutrient, plays a pivotal role in the plant growth systems. It is an integral part of the proteins, enzymes and nucleic acids which are responsible for the development of chlorophyll. Thus, nitrogen supply to plants is of utmost importance in all the crops.

Populus deltoides, locally called 'Fras', make a striking and important contribution to the landscape and economy of Jammu and Kashmir. Poplars have gained considerable importance in farm and plantation forestry in the state like the other neighbouring states of Uttarakhand, Haryana and Punjab. Poplars are fast growing trees; they recycle nutrients quickly due to their shedding of a large quantity of leaves which decompose early. Poplar timber is used for making apple boxes, interior wood work, beams, poles and fuelwood. Poplar is one of the few forest species which is considered ideal for successful intercultivation with agricultural crops. Poplars are known for their fast growth and easy vegetative propagation. They enrich the soil with litter and provide high production (10-30 m^3/ha/ year) on a short rotation of 8-12 years (Chandra,1986). The Food and Agriculture organization (FAO) has recommended the introduction of poplars to meet the increasing timber requirements of the world. Poplars have the potential to narrow the gap between the demand and supply of wood. Therefore, various attempts have been made in the past for raising this economically important species on problematic lands in different parts of the world.

In temperate regions, particularly in the Kashmir valley, farmers possess small land holdings, which are mostly under the cultivation of agricultural crops. Therefore, farmers are left with the only option of planting trees on degraded lands. As the availability of land for tree plantation is insufficient, people have to grow short rotation crops. Short rotation species are seen as an option to have better income and biomass in a short time and in a sustainable way (Bentsen and Felby, 2012).They are cultivated not only on arable lands, but also on problematic soils. The most important exotic species used for this purpose in the region is *Populus deltoides*. It is the most commonly cultivated species in community and household woodlots. In an environment suffering from land degradation and deterioration, the fast growing and resilient *Populus deltoides* performs better than most indigenous tree species. Small holders show a clear preference for this poplar which has multiple uses (Jagger and Pender, 2003). Therefore, an attempt has been made to utilise degraded lands for Poplar plantation to address the problem of land degradation with a social commitment.

The current emphasis on plantations has regularly been confined to merely achieving the targets without taking cognisance of the success of the plantations. Greening of degraded and problematic landscapes is the most important task of all afforestation activities. It is a common sight in the hills and plains to see the same area being planted and replanted, year after year, with a little or negligible success. A major reason behind this fiasco is the lack of technical know- how. The common practice prevailing for out planting results in seedling moisture stress and competition with weeds for light, moisture, nutrients and space. Moisture conservation is the quintessence of any plantation strategy on problematic sites for it to be successful. The use of appropriate soil working techniques and management ensures more infiltration and higher moisture retention in the soil.

Faulty land use systems and inflated human population have led to land degradation in the hills and plains. The efforts of the planners, foresters and scientists have brought a large chunk of areas under plantation, but the success rate is decreasing year after year. The reason is the prevalence of age- old practices and poor physicochemical attributes of such soils (problematic lands). In order to restore the productive potential of such lands, soil working and use of suitable trees, along with other measures like fertiliser application, moisture conservation etc., are alternatives for ameliorating these problematic lands. At present, the knowledge about the planting techniques on problematic sites is very limited and moreover, specific to specific areas. Therefore, there is an urgent need to undertake such work on scientific lines to develop techniques in tune with the existing problems.

Types of Degraded Lands in Temperate Regions

The causes of different types of land degradation are: water erosion, wind erosion, soil fertility decline, steepness of slope, barren rocky areas, salinisation, waterlogging and lowering of the water table.

Water erosion: It covers all forms of soil erosion by water, including sheet and rill erosion and gullying. It also includes human-induced intensification of land sliding caused by vegetation clearance, road construction etc.

Wind erosion: It refers to loss of soil by wind, occurring primarily in dry regions.

Soil fertility decline: It is used to briefly refer to what is more precisely described as deterioration in the soil's physical, chemical and biological properties. Whilst decline in fertility is indeed a major effect of erosion, the term is used here to refer to the effects of processes other than erosion. The main processes involved are:

- lowering of soil organic matter, with an associated decline in soil biological activity;
- degradation of soil physical properties (structure, aeration, water holding capacity), as brought about by reduced organic matter;
- adverse changes in soil nutrient resources, including reduction in the availability of the major nutrients (nitrogen, phosphorus, potassium), onset of micronutrient deficiencies and development of nutrient imbalances; and
- build-up of toxicities, primarily acidification, through incorrect fertiliser use.

Waterlogging: It is the lowering in land productivity through the rise in groundwater close to the soil surface. Also included under this heading is its severe form, termed ponding, where the water table rises above the surface. Waterlogging is linked with salinisation, both being brought about by incorrect irrigation management.

Salinization: The term is used in its broad sense to refer to all types of soil degradation brought about by the increase of salts in the soil. It thus covers both salinisation in its strict sense, the build-up of free salts and codification (also called alkalization), the development of the dominance of the exchange complex by sodium. As human-induced processes, these occur mainly through the incorrect planning and management of irrigation schemes. The term also covers saline intrusion, which is the incursion of sea water into coastal soils, arising from over- abstraction of groundwater.

Lowering of the water table: It is a self-explanatory form of land degradation, brought about by tube well pumping of groundwater for irrigation, in excess of the natural recharge capacity. This occurs in areas of non-saline ('sweet') groundwater. Pumping for urban and industrial use is another cause for the lowering of the water table.

Reversible Degradation and Land Reclamation

The effects of water and wind erosion are largely irreversible. Although plant nutrients and soil organic matter may be replaced, the replacement of the actual loss of soil material would require taking the soil out of use for many thousands of years, an impractical course of action.

In other cases, land degradation is reversible. Soils with reduced organic matter can be restored by the addition of plant residues and degraded pastures may recover under improved range management. Salinised soils can be restored to productive use, although at a high cost, through salinity control and reclamation projects.

Land reclamation frequently requires inputs which are costly, labour-demanding or both. The reclamation projects in salinised and waterlogged irrigated areas demonstrate this fact clearly. In other cases, the land can only be restored by taking it out of productive use for some years, as in reclamation forestry. The cost of reclamation or restoration to productive use of degraded soils is invariably less than the cost of preventing degradation before it occurs.

Plantation Strategy in Temperate Regions

Three types of pits are mostly used to establish and grow poplar plantations in degraded sites:

Types of Pits

Three types of pits are mostly used to establish and grow poplar plantation in degraded sites pertaining.

1. Ordinary pit (Auger planting 60 cm deep)
2. Saucer pit
3. Ring pit

Mulching Types

1. Ordinary mulch (grasses)
2. Black polythene sheet

Hydrogel Application on

1. Dry Application
2. Wet Application

Hydrogels

Water Jelly Crystals are an example of a water-absorbing polymer called a hydrogel. These hydrogels are long chains of molecules (called polymers) and absorb incredible amounts of water, only to release it to plant roots at a later time. Hydrogels are rapidly becoming one of the most exciting environmental education topics in classrooms worldwide. Today, superabsorbent polymers are widely used in such applications as forestry, gardening and landscaping as a means of conserving water. Imagine using a substance that could store water in the soil and then release it, as and when required by the roots of the plants. While we may consider water-absorbing polymers to be a modern convenience, imagine the impact that this technology is having on those parts of the world that are plagued by drought.

General Performance

One pound of Hydrogel will absorb up to 35 gallons of rainwater or snowmelt and 20-25 gallons of tap water, depending on the salt content of the water. Hydrogel can be applied wet or dry. Dry granules are usually easier to use, but they can be soaked thoroughly to fully fill them with water. When hydrated, the granules look like chunks of clear gelatin, about 1/2 inch in diameter.

Dry Application

For large quantities of potting soil or backfill around trees and shrubs, use 1.5-2 pounds/ cubic yard of potting soil or backfill or 1 ounce/ cubic foot of soil. For small quantities of potting soil, use 1/2 tsp per quart of soil.

Note: Since dry granules swell to many times their original size when water is added, 15-20% swelling room must be left in each planting hole or flower pot to compensate.

Wet Application

It is best for small applications such as repotting house plants, planting shrubs and small trees and bedding plants. 1/2 tsp of dry granules absorbs approximately

1 cup of water. 1 ounce of dry granules absorbs approximately 1 1/2 cups of water. 1 pound of dry granules absorbs approximately 30 gallons of water. Mix the granules in water and allow them to stand for 60-90 minutes (hot water works faster). Once the polymer is thoroughly soaked, the application rate is roughly one part hydrated polymer to four parts soil.

Hydrogel in Crystal Form

Case Study

Effect of different moisture conservation measures on moisture regimes of soils in poplar plantations done on problematic sites

The mulching materials (control, ordinary mulch and black polythene) were found to vary the moisture percentage under the poplar plantations done in autumn planting at problematic sites. Moisture percentage under black polythene was found to be the highest as compared to ordinary mulch and control in both planting seasons. The graphical representation categorised on the basis of fortnight intervals with moisture percentage showed that moisture remained static in interval P1 (1st-15th January) and P2 (16th - 31th January), increased drastically in P4 (16th -28th February), showed a declining trend up to P15 (1st-15th August) and then increased again. The highest percentage of moisture was found in P4 (16th-28th February) and lowest in P15 (1st-15th August) in autumn planting. In spring planting, the highest percentage of moisture was reported in P2 (16th -30th April) and lowest in P9.

The pits of different shapes were found to vary according to the availability of moisture percentage on the plantation site in both planting seasons. The graphical representation of fortnight intervals with moisture percentage showed that among the three pit types, saucer pit was recorded with the highest percentage of moisture, which was further followed by ring pit and ordinary pit. The moisture percentage was found to increase from P1 to P2 and thereafter, showed a declining trend. The highest percentage of moisture was observed in P2 (16th – 31st January) and lowest in P20 (16th-31st October) in autumn planting. However, in spring planting, the moisture percentage showed a decreasing trend. The highest moisture percentage was found in interval P1 (1st-15th April) and lowest in P14 (16th-31th October).

It was found that among the three pit sizes, the highest moisture was conserved under 60×60 cm^3 followed by 45×60 cm^3 and 30×30 cm^3 in both planting seasons. The percentage of moisture was recorded after every 15 days interval and it was found that the highest percentage was observed in P1 (1st- 15th January) and lowest in P20 (15th-31st October). In general, moisture percentage showed a declining trend from P1 – P15 in autumn planting . It was observed that in spring planting, the highest moisture percentage was recorded in P1 (1st -15 January) and P3 (1st-15 February) and the lowest in P14 (15th-31th July).

Effect of different mulching materials on moisture regimes of soils in poplar plantations done on problematic sites in autumn season

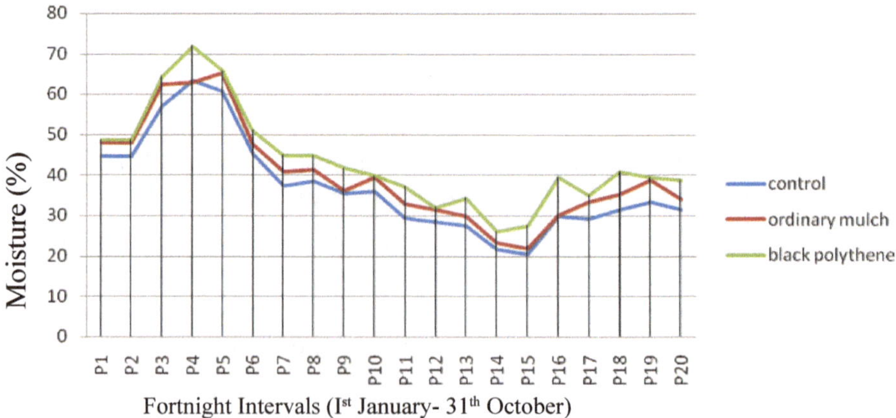

Effect of different pit shapes on moisture regimes of soils in poplar plantations done on problematic sites in autumn season

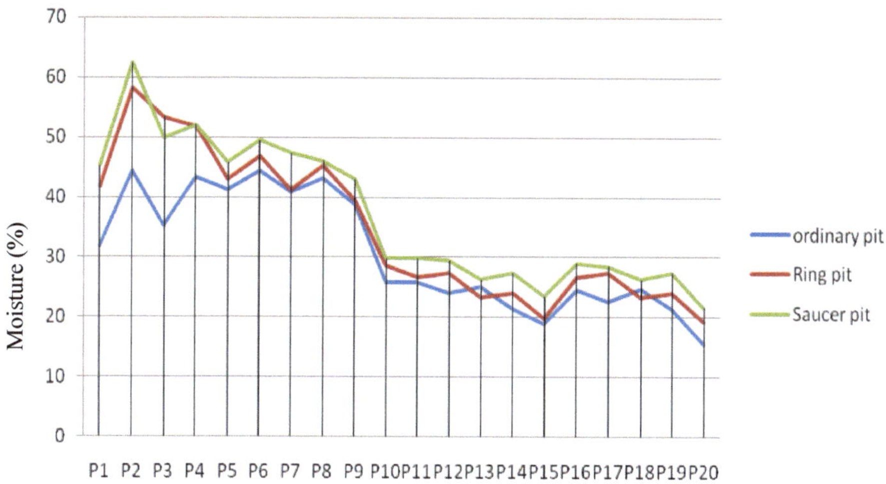

Effect of different pit sizes on moisture regimes of soils in poplar plantations done on problematic sites in autumn season

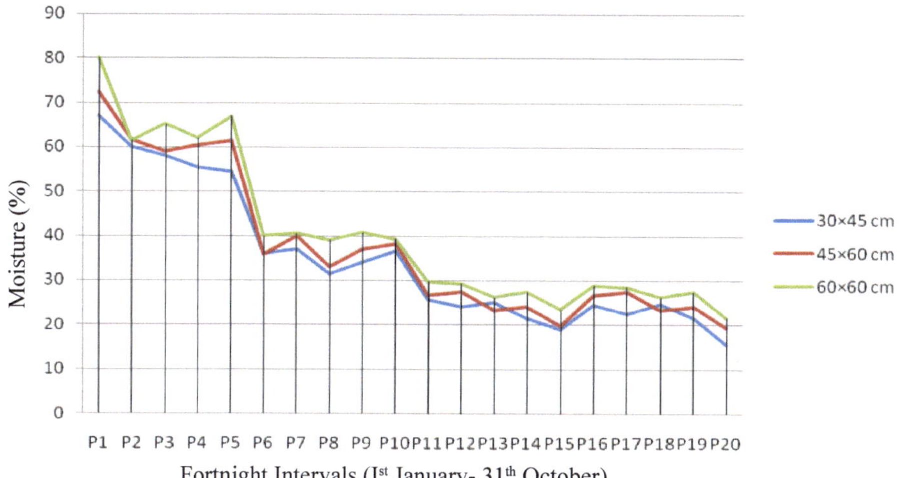

Effect of different mulching materials on moisture regimes of soils in poplar plantations done on problematic sites in spring season

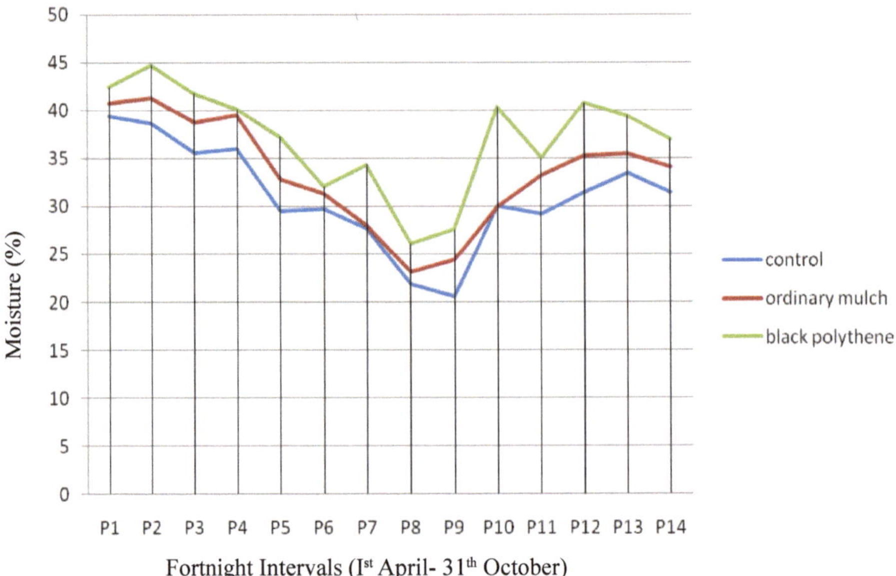

Effect of different mulching materials on moisture regimes of soils in poplar plantations done on problematic sites in spring season

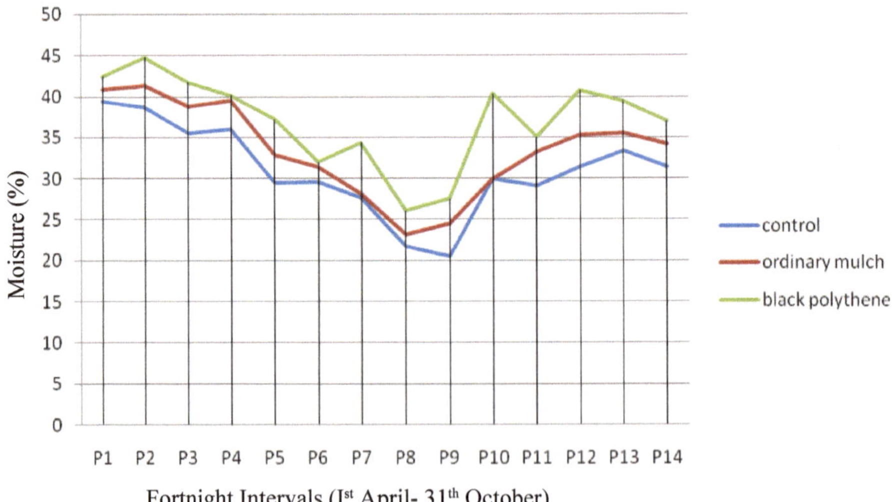

Effect of different pit shapes on moisture regimes of soils in poplar plantations done on problematic sites in spring season

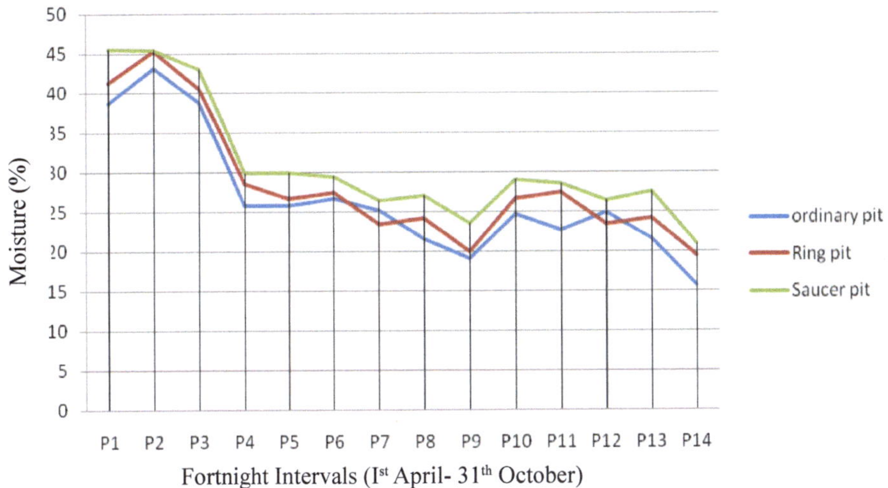

Effect of different pit sizes on moisture regimes of soils in poplar plantations done on problematic sites in spring season

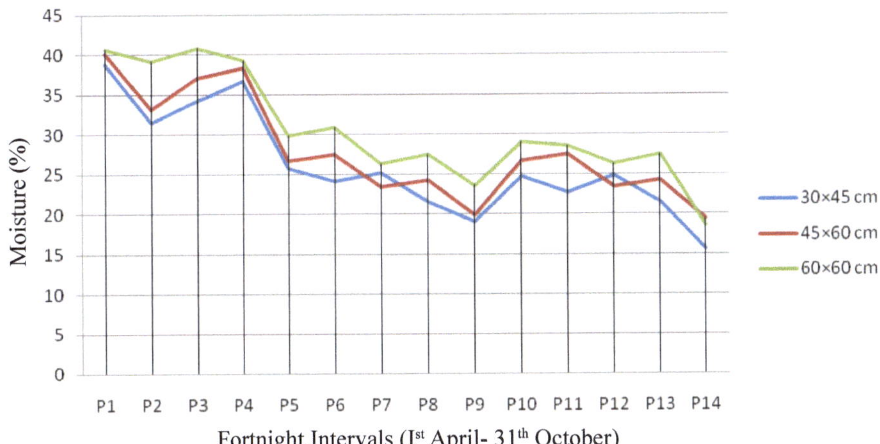

Relative Economics of Out Planting Operations

An economic analysis of *Populus deltoides* plantation on problematic sites in the temperate regions of the north-western Himalayas was carried out in the Faculty of Forestry, Sher-e-Kashmir University of Agricultural Sciences and Technology of Kashmir, Jammu and Kashmir. The actual data for cost and benefit were obtained for one growing season and the expected data was calculated after consulting economists and timber traders to obtain the total benefits for an eight-year rotation cycle. Profitability analysis of plantation was carried out to calculate the economics of *Populus deltoides* on problematic sites. The total plantation area was 0.25 hectare (5 kanals). The average cost of plantation was found to be Rs. 1,10,361.00 for 8 years, while the gross return was Rs. 9,32,300.00. The total variable cost was found to be Rs 97,086.00, which includes the cost of labour, fertilisers, planting material, mulching material and transportation. The maximum expenditure incurred on planting material was Rs. 30,000.00 (30.90 %). Planting operations and transportation costs claimed Rs. 20,000.00 (20.60 %). The cost of labour and felling of branch wood and timber was Rs. 38,000.00 (39.14 %). while fertilisers and mulching materials cost Rs. 5000.00 (5.15 %). The minimum expense was incurred on (purchase and maintenance of miscellaneous expenditures Rs. 4086.00) (4.20%). The total fixed cost was Rs. 13,275.00, in which, land rent and the depreciation cost of implements was included. The amount claimed by land rent was Rs. 11,500.00 (86.62 %) and the depreciation cost of implements was Rs. 17,275.00 (13.38 %). The net return (GR-TC) of the plantation was calculated and found to be Rs 8,21,939.00. The maximum income, i.e., Rs. 7,90,000.00 (96.11 %) was generated by timber in an 8- year rotation cycle. Fuel wood and branch wood contributed Rs. 31,939.00 (3.89 %). The return over variable cost (GR-TVC) and return over fixed cost (GR-TFC) were also calculated and were found to be Rs. 8,35,214.00 and Rs 9,19,025.00 respectively. The Net Present Value (NPV) of the total cost (@ 10 %) and Net Present Value (NPV) of the return (@ 10 %) were estimated and found to be Rs 72,071.54 and Rs 4,37,488.88 respectively. The net present worth of plantation was Rs 3,65,417.30. The estimated benefit- cost ratio of *Populus deltoides* plantations done on problematic sites might be Rs 6.07.

Table 1. Estimated cost and return of *Populus deltoides* on problematic sites of Temperate Regions

	Years of Production							
Particulars (in Rupees)	Ist	2nd	3rd	4th	5th	6th	7th	8th
Total variable cost	34936	450	700	0	7500	5000	8000	40500
Total fixed cost	2090	1580	1605	1500	1500	1690	1500	1810
Total cost	37026	2030	2305	1500	9000	6690	9500	42310
Production (qt.)	0	0	25	0	50	65	95	4544(ft.)
Gross return	0	0	2500	0	5000	6500	9500	908800

Table 4. Profitability Analyses

S. No.	Particulars	Amount (in Rupees)
1.	Total variable cost	97086
2.	Total fixed cost	13275
3.	Total cost	110361*
4.	Gross return	932300*
5.	Net return (GR-TC)	821939
6.	Return over variable cost (GR-TVC)	835214
7.	Return over fixed cost (GR-TFC)	919025
8.	Net present value of total cost (@ 10 %)	72071.54
9.	Net present value of return (@ 10 %)	437488.88
10.	Benefit of return (B/C)	6.07
11.	Net present worth	365417.3

View of Degraded Site

Plantation Site After one Growing Season

References

Banerjee, B and Raza, M. 1992. Strategies for restoration of wastelands in India. Development and ecology essays. Rawat Publications, Jaipur, India.

Bentsen, N. S. and Felby, C. 2012. Biomass for energy in the European union, a review ofBioenergy resource assessment. Biotechnology Biofuels, 5 (1): 25.

Bremer, L. and Farley, K. 2010. Does plantation forestry restore biodiversity or create green deserts? A synthesis of the effects of landuse transitions on plant species richness. Biodiversity Conservation, 19: 3893-3915.

Chandra, J.P. 1986. Poplars a cash crop for north Indian farmers. Indian Forester 112(8): 698-710.

Digest of forest statistics, Jammu and Kashmir, 2012. Pp. 29.

Jagger, P. and Pender, J. 2003. The role of trees on sustainable management of less favored lands: the case of eucalyptus in Ethiopia. Forest policy and economics 5: 83-95.

Jin, Z. Z. 2002. Floristic features of dry hot and dry warm valleys, Yunnan and Sichuan, YunnanScience and Technology Press, Kunming, China.

Moscatelli, M. C., Lagomarsino, A. P. D. and Grego, S. 2008. Short and medium term contrasting effects of nitrogen fertilization on C and N cycling in a poplar plantation. Forest ecology and management 255: 447-454.

Pramod, J. and Mohapatra, K. P. 2012. Devoloping a soil quality index for ravenous ustifluvents of the semi-arid region of India. Journal of the Indian Society of Soil Science 60 (2): 116-124.

Thapa, R. 2003. Agroforestry can reverse land degradation in Nepal. Appropriate Technology(30): 40-41.

Tomar, O. S.,Minhas, P. S., Sharma, V. K., Singh, Y.P. and Gupta, R.K. 2003. Performance of 31 tree species and soil conditions in a plantation established with saline irrigation. Forest ecology and Management 177: 333-346.

15

Management of Abiotic Stress in Forest Nurseries

Z.M. Dar

Division of Basic Sciences and Humanities, FoA, Wadura, Jammu and Kashmir, India

Introduction

Forest trees constitute ~82% of the continental biomass and harbour >50% of the terrestrial biodiversity. They also help to mitigate climate change, and provide a wide range of products that meet human needs, including wood, biomass, paper, fuel and biomaterials. Field-grown trees are routinely exposed to environmental stress. Current and predicted climatic conditions, such as prolonged drought, increased salinisation of soil and water and low temperature episodes pose a serious threat to forest productivity worldwide, affecting tree growth and survival. According to the reports of the Intergovernmental Panel on Climate Change, these stresses will increase in the near future because of global climate change. Hence, the need of the hour is to formulate strategies that could alleviate the adverse effects of abiotic stresses on forest species. This is not possible unless and until we have knowledge of their effect on the various physiological aspects of the plant. Hence, an attempt has been made in the following sections of the paper to discuss:

- The adverse effects of abiotic stresses on plants, the most prominent among them being drought and high temperature in the Kashmir Valley.
- The practices that could mitigate the harmful effects of drought and high temperature in plants, especially at the nursery stage.

Effects of Drought Stress

Drought is considered the single most devastating environmental stress, which decreases crop productivity more than any other environmental stress (Lambers *et al.*, 2008). A continuous shortfall in precipitation coupled with a higher evapotranspiration demand leads to drought, which is defined as the lack of ample moisture required for normal plant growth and development to complete the life cycle. In plants, a series of biochemical, physiological and morphological injuries occur due to water stress, depending on factors like the duration of exposure of plants to the drought, genetic resistance and stage of growth.

Effect on Growth

The first and foremost effect of drought is impaired germination and poor stand establishment (Harris *et al.*, 2002). Drought stress has been reported to severely reduce germination and seedling stand. Growth is accomplished through cell division, cell enlargement and differentiation and involves genetic, physiological, ecological and morphological events and their complex interactions. The quality and quantity of plant growth depend on these events, which are adversely affected by water deficit. Cell growth is one of the most drought- sensitive physiological processes due to the reduction in turgor pressure (Taizand Zeiger, 2006). Due to water deficiency, cell elongation of higher plants can be inhibited by the interruption of water flow from the xylem to the surrounding elongating cells (Nonami, 1998), thus restricting plant growth.

Effect on Photosynthesis

A major effect of drought is the reduction in photosynthesis, which arises by a decrease in leaf expansion, impaired photosynthetic machinery, premature leaf senescence and the associated reduction in food production (Wahid and Rasul,2005). When stomatal and non- stomatal limitations to photosynthesis are compared, the former can be quite small. This implies that other processes,besides CO uptake, are being damaged. The role of drought- induced stomatal closure, which limits CO uptake by leaves, is very important. In such events, restricted CO availability could possibly lead to increased susceptibility to photo-damage (Cornic and Massacci, 1996). Drought stress produced changes in the photosynthetic pigments and their components (Anjum *et al.*, 2003), damaged photosynthetic apparatus (Fu J. and Huang, 2001) and diminished the activities of Calvin cycle enzymes, which are important causes of reduced crop yield (Monakhova and Chernyadèv, 2002). Another important effect that inhibits the growth and photosynthetic abilities of plants is the loss of balance between the production of reactive oxygen species and the antioxidant defence (Reddy *et al.*, 2004), causing the accumulation of reactive oxygen species, which induces oxidative stress in proteins, membrane lipids and other cellular components.

Assimilate Partitioning

Assimilate translocation to reproductive sinks is vital for seed development. Seed set and filling can be limited by the availability or utilisation i.e., assimilate source or sink limitation, respectively (Asch *et al.*, 2005). Drought stress frequently enhances the allocation of dry matter to the roots, which can, in turn, enhance water uptake (Leport *et al.*, 2006). Drought stress decreases the photosynthetic rate and disrupts the carbohydrate metabolism and level of sucrose in leaves that spills over to a decreased export rate. This is presumably due to drought stress-induced increased activity of acid invertase (Kim *et al.*, 2000). Limited photosynthesis and sucrose accumulation in the leaves may hamper the rate of sucrose export to the sink organs and ultimately, affect the reproductive development.

Apart from source limitation, the capacity of the reproductive sinks to utilise the incoming assimilates is also affected under drought stress and may also play a role in regulating reproductive abortion. Drought-induced carbohydrate deprivation, enhanced endogenous abscisic acid concentration and an impaired ability of the reproductive sinks to utilise the incoming sucrose are potential factors contributing to seed abortion in grain crops. A reduced acid invertase activity can arrest the development of reproductive tissues due to improper phloem unloading (Goetz *et al.*, 2001). In addition, drought stress may inhibit the important functions of vacuolar invertase mediated sucrose hydrolysis and osmotic potential modulation. In drought-stressed maize, a low invertase activity in the young ovaries lowers the ratio of hexose to sucrose. This may inhibit cell division in the developing embryo/endosperm, resulting in weak sink intensity and may ultimately lead to fruit abortion (Andersen *et al.*, 2002).

In summary, drought stress not only limits the size of the source and sink tissues, but phloem loading, assimilate translocation and dry matter portioning are also impaired. However, the extent of these effects varies with the plant species, stage, duration and severity of drought.

Oxidative Damage

The exposure of plants to certain environmental stresses quite often leads to the generation of reactive oxygen species, including superoxide anion radicals (O− 2), hydroxyl radicals (OH), hydrogen peroxide ($H2O^2$), alkoxy radicals (RO) and singlet oxygen (O1 2) (Munné-Bosch and Penuelas, 2003). Reactive oxygen species may react with proteins, lipids and deoxyribonucleic acid, causing oxidative damage and impairing the normal functions of cells (Foyer and Fletcher, 2001). Many cell compartments produce reactive oxygen species. Of these, chloroplasts are a potentially important source because excited pigments in thylakoid membranes may interact with O^2 to form strong oxidants such as O^{-2} or O^{-1}, OH^- and H_2O_2 (Reddy etal.,2004). The interaction of O_2 with the reduced components of the electron transport chain in mitochondria can lead to reactive oxygen species formation (Moller, 2001).

Effects of Heat Stress

High temperature (HT) stress is a major environmental stress that limits plant growth, metabolism and productivity worldwide. Plant growth and development involve numerous biochemical reactions that are sensitive to temperature. Heat stress is often defined as the rise in temperature beyond a threshold level for a period of time sufficient to cause irreversible damage to plant growth and development. In general, a transient elevation in temperature, usually 10–15 °C above ambient, is considered heat shock or heat stress. However, heat stress is a complex function of intensity (temperature in degrees), duration and rate of increase in temperature. Heat stress due to high ambient temperatures is a serious threat to crop production worldwide. Gaseous emissions due to human activities are substantially adding to the existing concentrations of greenhouse gases, particularly CO , methane, chlorofluorocarbons and nitrous oxides. Different global circulation models predict that greenhouse gases will gradually increase the world's average ambient temperature. According to a report of the Intergovernmental Panel on Climatic Change (IPCC), the global mean temperature will rise 0.3 °C per decade (Jones *et al.*, 1999), reaching to approximately 1 and 3 °C above the present value by the years 2025 and 2100, respectively, thus leading to global warming. Rising temperatures may lead to altered geographical distribution and growing season of agricultural crops by allowing the threshold temperature for the start of the season and crop maturity to be reached earlier (Porter, 2005). Plant responses to HT vary with the degree of temperature, duration and plant type. At extreme HT, cellular damage or cell death may occur within minutes, which may lead to a catastrophic collapse of cellular organisation. Heat stress affects all aspects of plant processes like germination, growth, development, reproduction and yield. Heat stress differentially affects the stability of various proteins, membranes, RNA species and cytoskeleton structures and alters the efficiency of enzymatic reactions in the cell, which hinders major physiological processes, thus creating metabolic imbalance.

Effect on Growth

Among the growth stages of a plant, germination is the first to be affected by heat stress. Heat stress has a negative impact on various crops during seed germination, though the range of temperature may vary largely according to the crop species. Reduced germination percentage and plant emergence, abnormal seedlings, poor seedling vigour and reduced radicle and plumule growth of geminated seedlings are major effects of heat stress documented in various cultivated plant species. High temperature causes loss of cell water content, due to which, the cell size and ultimately, the growth is reduced. The reduction in net assimilation rate (NAR) is also another reason for reduced relative growth rate (RGR) under HT. The morphological symptoms of heat stress include scorching and sunburn of leaves, twigs, branches and stems, leaf senescence and abscission, shoot and root growth inhibition, fruit discoloration and damage.

Effect on Photosynthesis

Photosynthesis is one of the most heat sensitive physiological processes in plants. High temperature has a greater influence on the photosynthetic capacity of C3 plants than C4 plants. In chloroplasts, carbon metabolism of the stroma and photochemical reactions in thylakoid lamellae are considered as the primary sites of injury due to HT. The thylakoid membrane is highly susceptible to HT. Major alterations occur in chloroplasts, like altered structural organisation of thylakoids, loss of grana stacking and swelling of grana under heat stress. The photosystem II (PSII) activity is also greatly reduced or even stopped under HT. Heat shock reduces the amount of photosynthetic pigments. The ability of the plant to sustain leaf gas exchange and CO^2 assimilation rates under heat stress is directly correlated with heat tolerance. Heat markedly affects the leaf water status, leaf stomatal conductance (gs) and intercellular CO^2 concentration. The closure of stomata under HT is another reason for impaired photosynthesis that affects the intercellular CO^2. The decline in chlorophyll pigment is also the result of lipid peroxidation of chloroplast and thylakoid membranes due to heat stress (40/30° C, day/night), as observed in sorghum. Photosystem II photochemistry (Fv/Fm ratio) and gs were also reduced under the same stress conditions. Other reasons believed to hamper photosynthesis under heat stress include the reduction of soluble proteins, Rubisco binding proteins (RBP), large subunits (LS) and small subunits (SS) of Rubisco in darkness and increases of those in light. High temperature also greatly affects starch and sucrose synthesis, by reduced activity of sucrose phosphate synthase, ADP-glucose pyrophosphorylase and invertase. Heat has negative effects on the leaf of the plant like reduced leaf water potential, reduced leaf area and premature leaf senescence, which have a negative impact on the total photosynthesis performance of the plant. Under prolonged heat stress, depletion of carbohydrate reserves and plant starvation are also observed.

Effect on Membrane Stability

The stability of various cellular membranes is important during high-temperature stress, just as it is during chilling and freezing. The excessive fluidity of membrane lipids at high temperatures is correlated with a loss of physiological function. In oleander (*Nerium oleander*), acclimation to high temperatures is associated with a greater degree of saturation of fatty acids in membrane lipids, which makes the membranes less fluid. At high temperatures, there is a decrease in the strength of hydrogen bonds and electrostatic interactions between polar groups of proteins within the aqueous phase of the membrane. High temperatures thus modify membrane composition and structure and can cause the leakage of ions. Membrane disruption also causes the inhibition of processes, such as photosynthesis and respiration that depend on the activity of membrane-associated electron carriers and enzymes. Photosynthesis is especially sensitive to high temperature. In their study of *Atriplex* and *Tidestromia*, O. Bjorkman and his colleagues (1980) found that electron transport in photosystem II was more sensitive to high temperature in the cold-adapted *A. sabulosa* than in the heat-adapted *T. oblongifolia*. In these plants, the enzymes ribulose-1,5-bisphosphate carboxylase, NADP: glyceraldehyde-3-phosphate dehydrogenase and phosphoenolpyruvate carboxylase were less stable at high temperatures in *A. sabulosa* than in *T. oblongifolia*. However, the temperatures at which these enzymes began to denature and lose activity were distinctly higher than the temperatures at which photosynthesis began to decline. These results suggest that the early stages of heat injury to photosynthesis are more directly related to the changes in membrane properties and to the uncoupling of the energy transfer mechanisms in chloroplasts than to a general denaturation of proteins.

Management Practices to Overcome the Adverse Effects of Drought and High Temperature

Agronomic Practices

The implementation of management practices can potentially alleviate the harmful effects of drought and heat stresses and include soil management and cultural practices, irrigation, crop residues and mulching and selection of more appropriate crop varieties.

Nutrient Application

Under heat stress, the application of macronutrients, such as K, Ca and micro-nutrients like B, Se and Mn, which are known to modify stomatal function, can help activate the physiological and metabolic processes contributing to pre-serving high water potential in tissues, thereby increasing heat stress tolerance (Waraich *et al.*, 2012). The application of nutrients such as N, K, Ca and Mg was also reported to reduce toxicity to ROS by increasing the concentration of antioxidant enzymes in plant cells (Waraich *et al.*, 2012). On the other hand, several studies have shown that the application of fertiliser has no significant effect on drought stress and optimal soil moisture content is required since water is critical for the mobility and metabolism of these nutrients (Lipiec *et al.*, 2013).

Application of Growth Regulators

Attention has also been focused on the application of plant growth regulators known to be involved in the response to stress. Among the plant growth substances, salicylic acid, cytokinins and ABA have been reported to play a key role in drought tolerance. Under water stress conditions, plant growth regulator treatments can significantly increase the water potential and chlorophyll content. The exogenous application of ABA increased soybean yields under water deficit conditions. New ABA formulations are currently available for commercial growers to delay drought-induced wilting symptoms and improve drought tolerance (Sharma *et al.*, 2006). Recent research has investigated the use of concentrated ABA or ABA analogs to maintain the marketability of horticulture crops by reducing drought stress symptoms (Blanchard *et al.*, 2007). ABA application during spring or summer also reduces transpiration in potted miniature rose (*Rose hybrida* L.) and results in better flower longevity.

Microbial Approaches

The ever- changing environmental conditions have been faced by plants through evolutionary time, among which, drought is considered as the most common, having an adverse effect on the growth and development of the plants, which is one of the major constraints on plant productivity worldwide and is expected to increase with climatic changes. The symbiotic relationship between arbuscular mycorrhizal (AM) fungi and the roots of higher plants is widespread in nature. Several ecophysiological studies have demonstrated that AM symbiosis is a key component in helping plants to cope with water stress and in increasing drought resistance, by bringing about various changes in the host plant at the morphological level, like alterations in root morphology to form a direct pathway of water uptake by extra radical hyphae. The physiological mechanisms promoting enhanced drought tolerance of the host plant by AM inoculation include increased photosynthetic rate and nutrient mobilisation followed by their rapid uptake under drought conditions. The biochemical mechanisms promoting drought tolerance in AM inoculated plants include the accumulation of osmoprotectants like proline, sugars and trehalose and a rapid enhancement in the concentration of different enzymatic and non-enzymatic antioxidants which reduce the risk of free radical attack on the plant cell membrane. which has been evidenced in terms of reduced electrolyte leakage and malondialdehyde (MDA) content of drought stressed plant tissues in various studies (Zaffar *et al.*, 2018).

References

Andersen M.N. Asch F. Wu Y. Jensen C.R..Næsted H., Mogensen V.O. Koch K.E. (2002). Soluble invertase expression is an early target of drought stress during the critical, abortion–sensitive phase of young ovary development in maize, Plant Physiol. 130:591–604.

Anjum, F. M. Yaseen E. Rasul A. Wahid S. Anjum (2003a). Water stress in barley (*Hordeum vulgare* L.). I. Effect on morphological characters. Pakistan J. Agric. Sci. 40: 43–44.

Asch F. Dingkuhnb M. Sow A. Audebert A. (2005). Drought-induced changes in rooting patterns and assimilate partitioning between root and shoot in upland rice, Field Crop. Res. 93: 223–236.

Bjorkman O. Badger, M. R..Armond P. A. (1980). Response and adaptation of photosynthesis to high temperatures. In: Turner, N. C.; Kramer, P. J. Adaptation of plants to water and high temperature stress. New York: J. Wiley & Sons,. p. 233-249.

Blanchard M.G. Newton L.A. Runkle E.S.D. Woolard. (2007). Exogenous applications of abscisic acid improved the postharvest drought tolerance of several annual bedding plants. Acta Hort. 755:127–132.

Cornic G. Massacci, A. (1996). Leaf photosynthesis under drought stress, in: Baker N.R., (Ed.), Photosynthesis and the Environment, Kluwer Academic Publishers, The Netherlands.

Fazeli F. Ghorbanli M..Niknam V. (2007).Effect of drought on biomass, protein content, lipid peroxidation and antioxidant enzymes in two sesame cultivars, Biol. Plant. 51: 98–103.

Foyer C.H. Fletcher J.M. (2001). Plant antioxidants: colour me healthy. Biologist. 48: 115-120. Fu J. Huang B. (2001) Involvement of antioxidants and lipid peroxidation in the adaptation of two cool-season grasses to localized drought stress. Environ. Exp. Bot. 45: 105–114. Goetz M. Godt D.E. Guivarch A., Kahmann U. Chriqui D. Roitsch T. (2001) Induction of male sterility in plants by metabolic engineering of the carbohydrate supply. Proc. Natl Acad. Sci. (USA) 98, 6522–6527

Hall, A. E. (2001). Crop Responses to Environment. CRC Press LLC, Boca Raton, Florida Handbook of Photosynthesis, 2nd ed. CRC Press, Florida, Pp. 479–497

Harris D. Tripathi R.S. Joshi A. (2002) On-farm seed priming to improve crop establishment and yield in dry direct-seeded rice, in: Pandey S., Mortimer M., Wade L., Tuong T.P., Lopes K., Hardy B. (Eds.), Direct seeding: Research Strategies and Opportunities, International Research Institute, Manila, Philippines, pp. 231–240.

Jones H. G. (1999). Use of thermography for quantitative studies of spatial and temporal variation of stomatal conductance over leaf surfaces. Plant, Cell & Environ. 22: 1043-1055.

Kim J.Y. Mahé A. Brangeon J. Prioul J.L. (2000) A maize vacuolurinvertase, IVR2, is induced by water stress. Organ/tissue specificity and diurnal modulation of expression, Plant Physiol. 124: 71–84.

Lambers H. Chapin F. Pons T. (2008) Plant physiological ecology. Springer, New York, p 540

Leport L, Turner N.C. French R.J. Barr M.D. Duda R. Davies S.L. (2006) Physiological responses of chickpea genotypes to terminal drought in a Mediterranean-type environment, Eur. J. Agron. 11: 279–291.

Lipiec, J. Doussan, C. Nosalewicz, A. Kondracka, K. (2013). Effect of drought and heat stress on plant growth and yield: a review. Intern.Agrophys. 27: 463-477.

Moller I. M. (2001) Plant mitochondria and oxidative stress: electron transport, NADPH turnover, and metabolism of reactive oxygen species, Annu. Rev. Plant Phys. 52: 561-591.

Monakhova, O. F. Chernyadèv I. I. 2002. Protective role of kartolin-4 in wheat plants exposed to soil drought. Appl. Biochem. Micro. 38: 373–380

Munné-Bosch S., Penuelas J. (2003) Photo and antioxidative protection, and a role for salicylicacid during drought and recovery in field grown Philly reaangustifolia plants, Planta 217: 758–766.

Nonami H (1998). Plant water relations and control of cell elongation at low water potentials.J. Plant Res. 111: 373-382

Porter, J. R. (2005). Rising temperatures are likely to reduce crop yields. Nature: 436, 174

Reddy A.R..Chaitanya K.V. Vivekanandan M. (2004) Drought-induced responses of photosynthesis and antioxidant metabolism in higher plants, J. Plant Physiol. 161:1189–1202

Sharma P, Jha A. B. Dubey R. S. Pessarakli M. (2012). Reactive oxygen species, oxidative damage, and antioxidative defense mechanism in plants under stressful conditions. J Bot. 1–26.

Taiz L., Zeiger E. (2006) Plant Physiology, 4th Ed., Sinauer Associates Inc. Publishers, Massachusetts

Wahid, A.Rasul E. 2005. Photosynthesis in leaf, stem, flower and fruit. In: Pessarakli M. (Ed.),

Waraich E.A. Ahmad R..Halim A. Aziz T. (2012) Alleviation of temperature stress by nutrient management in crop plants: A review. J. Soil Sci. Plant Nutr. 12: 221–244.

Zaffar M. D. Amjad M. Malik A. Mushtaq A. M. (2018). Review on Arbuscular Mycorrhizal Fungi: An Approach to Overcome Drought Adversities in Plants. Inter. J. of Curr. Microbiol.and Applied Sci. 7 (03): 1040-1049

16

Management of Insect Pests of Forest Nurseries

Munazah Yaqoob, Quratul Ain, Sajad H. Mir and Liyaqat Ayub

Division of Entomology, Faculty of Agriculture, SKUAST-K, Wadura, Jammu and Kashmir, India

Introduction

A forest is a complex ecosystem, predominantly composed of trees and shrubs, which usually form a closed canopy. They form a storehouse of a large variety of life forms such as plants, mammals, birds, insects, reptiles etc., besides being the most important habitat of the microorganisms and fungi, which do the important work of decomposing dead organic matter, thereby enriching the soil. Nearly 4 billion hectares of forests cover the earth's surface, roughly 30% of its total land area. The forest ecosystem has two components- abiotic and biotic. Climate and soil type form a part of the non-living component and the living component includes plants, animals and other life forms. Plants include the trees, shrubs, climbers, grasses and herbs in the forest. Depending on the physical, geographical, climatic and ecological factors, there are different types of forests, like evergreen forests (mainly composed of evergreen tree species, i.e., species bearing leaves throughout the year) and deciduous forests (mainly composed of deciduous tree species, i.e., species having leaf fall during particular months of the year). Each forest type forms a habitat for a specific community of animals that are adapted to live in it. The term 'forest' implies the 'natural vegetation' of the area, existing from thousands of years and supporting a variety of biodiversity, forming a complex ecosystem. Plantation is different from natural forest as these planted species are often of the same type and do not support a variety of natural biodiversity. Forests provide various natural services and products. Many forest products are used in day-to-day life. Apart from this, forests play an important role in maintaining the ecological balance and contribute to the economy too. Forests provide a suitable environment for many species of plants and animals and thus, protect and sustain the diversity of nature.

India's forest cover is estimated to be about 67.701 million hectares or 22.8% of the country's total land area, with 3.226 million ha of forest plantations representing 4.8% of the total forest area (FAO, 2006).

Insect Pests of Forests

A large number of insects and diseases are known to damage both naturally regenerating forests and plantations in India, although very little data is available on the area affected by these insects. However, according to an estimate of FAO, 10,00,000 ha. of forest was damaged by insect pests (FAO, 2005). Herbivorous insects feed on different parts of the trees, such as leaves, wood, bark, inflorescence, roots etc., thus causing massive damage of tree health as well as timber quality. The wood and bark- boring insects, mainly belonging to the orders *Dictyoptera, Isoptera, Coleoptera, Lepidoptera* and *Hymenoptera*, bore into the wood in search of food or for shelter. Defoliators, on the other hand, skeletonise the plant, whereas sap suckers suck the sap of the leaves of plants belonging to *Hemiptera* and *Thysanoptera* orders (Mohandas *et al.*, 1990 and Singh *et al.*, 2005). As a result, the surface area for photosynthesis and transpiration is greatly reduced and the growth rate of trees as well as timber quality is also reduced. The temperate forests of Jammu and Kashmir state are also vulnerable to attack by many insect pests which adversely affect their growth and productivity. Plants in the age group of 3-6 years and height class 3-5 m are reported to be more susceptible to pest attack and reveal up to 90% infestation, showing the relationship between infestation and the age of the tree (Cipiao *et al.*, 2009). Repeated attacks in the early years of plantation eventually lead to the death of the trees (Lim *et al.*, 2008).

Major insect pests of the forest areas of Kashmir include:

White grub

White grubs are soil-dwelling larvae of insects commonly known as 'May beetles' or 'June beetles', belonging to the family *Scarabaeidae* of *Coleoptera*. These grubs feed on herbaceous plant roots and other soil organic matter, but will also feed on the roots of woody plants, including all types of coniferous and hardwood seedlings in nursery settings. The grubs are the major pests feeding on the roots of *Pinus wallichiana, Cedrus deodara, Cupressus torulosa* and *Robinia pseudoacacia* and are known to cause damage in forest nurseries. Different species of white grub attack forest trees which include *Holotrichia longipennis, Brahmina sp., Melolontha furcicauda, Centonia sp., Anomala sp.* and *Oryctes sp.*

White grub damage is typically noticed from June through early fall, when formerly healthy seedlings become discoloured, wilt and die. Above ground symptoms may resemble those of drought injury. Heavily damaged seedlings can be pulled gently from the soil due to extensive root loss. Below ground, the taproot or lateral roots may be chewed off, girdled, gouged or debarked.

Management

- Identify potential problem areas by scouting for white grubs in the soil.
- The fumigation of seed beds can eliminate white grubs in the upper soil horizons, but overwintering larvae, which reside below the effective fumigation zone, can be managed by granular and liquid formulations of insecticides.
- Irrigating the soil before and after insecticide application may help to bring grubs nearer to the surface and move insecticide into the soil, respectively.
- The application of contact insecticides on the foliage of the trees and plants in the surroundings can be carried out for the management of adult insects.
- Shaking of trees helps to shed the adult insects, which are then trapped in the kerosene kept in some tubs. This is an effective management practice.
- The installation of light traps reduces the population of adult beetles

Wireworm *Agriotes* sp.

It belongs to the family *Elateridae* of *Coleoptera* and is one of the major pests that attack forest trees, particularly nursery plants. The larvae feed and bore in the roots of *Pinus wallichiana, Cedrus deodara, Cupressus torulosa* and *Robinia pseudoacasia*. The affected plants wilt and die if the roots suffer a large amount of damage.

Cutworm *Agrotis* sp.

It belongs to the family *Noctuidae* of *Lepidoptera.* The larvae of several moths (order *Lepidoptera*) occur incidentally in forest nurseries. The larvae feed at night and hide during the day and are often difficult to detect. The damage is generally confined to young, succulent seedlings before their stems become woody. Cutworms can also feed on foliage and often cut through the stems just above ground, leaving a short stump. The following are indicators of the presence of cutworms: (i) stems on which the needles have been consumed, (ii) sunken or depressed areas on the stems that look like fungus-caused lesions (these are old feeding sites), (iii) stems cut off below the soil so it appears that the seedling did not germinate and (iv) stems cut off at ground level.

Management

Surroundings should be kept weed-free to reduce egg-laying females, which are attracted to certain weeds. However, once cutworms are present, killing weeds may drive the larvae from the preferred weeds to seedlings. The installation of light traps reduces the population of adult moths and subsequent larval damage. Poison baits with natural attractants (such as apple pomace or bran) may be used to kill the larvae. Another management practice is spraying large areas of outbreak with insecticides.

Mole cricket, *Gryllotalpa* sp.

It belongs to the family *Gryllotalpidae* of order *Orthoptera.* Nymphs and adults occasionally cause damage in forest areas. The damage is confined to newly planted seedlings in shaded areas. The crickets may feed aboveground on foliage or stem tissue and belowground on roots. Girdling of the stems of seedling plants at the soil surface is a common form of injury, though young plants are sometimes severed and pulled belowground to be consumed. Additional injury to small plants is caused by soil surface tunneling, which may dislodge seedlings or uproot them, leading to their desiccation.

Management

Liquid and granular formulations of insecticides are commonly applied to the soil to suppress mole crickets. In some cases, insecticide application should be followed by irrigation because the insecticide must enter the root zone of the plants to be most effective. Bait formulations are also useful and various baits containing wheat bran, cottonseed meal or some other grain product plus 2-5% toxicant have proven to be effective. The addition of 5 to 15% water and 2 to 5% molasses to the grain-toxicant mixture is sometimes recommended. The biological control of mole crickets can be enhanced by the application of the entomopathogenic nematode, *Steinernema sp.*

Green stink bug, *Nezaravirudula*

These bugs feed by sucking sap from plants. They are believed to inject a toxic substance into the plant when feeding to break down plant tissues. Their feeding is very destructive to fruit and other tender plant parts. They feed on swollen fruit and leaf buds, causing the buds to dry up.

Management

Preventing the growth of broad-leaved winter annual weeds and legumes in and around nursery areas can reduce the populations of these bugs. Spraying neem oil, pyrethoids and synthetic pyrethoids (Cypermethrin) is effective in controlling the bugs.

Aphids

Seedlings of all tree species are subject to attack by one or more species of aphids. This species of the genus *Cinara*, which attacks conifers, is probably the most important of the seedling-infesting aphids. Aphids feed on plant sap, causing a general weakness. Heavy feeding, however, may result in branch dieback or even the death of seedlings. Aphids also serve as vectors of plant pathogenic viruses which cause substantial damage. Also, aphids excrete honeydew, a sugary substance on which, a fungus called sooty mold grows, making the foliage appear black (Dumroese, 2012).

Management

Cultural - Maintain seedling vigour through proper fertilisation and irrigation to increase tolerance to aphids.

Chemical - The use of systemic insecticides, such as dimethoate, effectively controls aphids. Sprays should be applied to the point of runoff on all foliage and stem surfaces.

Sawflies

Several sawfly species (families: *Diprionidae* and *Tenthredinidae* of order *Hymenoptera*) are common pests of young conifers, but they rarely cause seedling mortality. Sawflies are divided into two groups: spring and summer sawflies. Spring sawflies generally feed on older foliage and summer sawflies feed on both old and new foliage. Defoliation caused by sawflies is generally light. The summer sawflies are the most destructive. They feed gregariously in small groups. Newly hatched larvae will often feed on needle edges. The damaged needles turn brown and sometimes, curl. As the larvae grow larger, they begin to consume the entire needle (Morris and Hoffard, 1989).

Management

Cultural: For localized infestations, hand removal of sawflies is done.

Chemical: Contact insecticides can be sprayed for controlling the larvae on a large scale. However, insecticide applications should target early stage larvae.

Leaf beetles

They belong to family *Chrysomelidae* of order *Coleoptera.* These beetles feed on the foliage and young succulent twigs of all plants, including poplar, giving a ragged foliage appearance. Some leaves will have brown patches where young larvae have skeletonised the leaves. Other leaves will have only their veins and midribs remaining. Heavy damage results in dead, black terminals with most of the leaf tissue consumed. Continuous defoliation and twig damage through the summer reduces seedling growth and vigour. Lateral buds sprout below the injured terminals and grow rapidly, resulting in multiple-forked tops. Stunted growth in nursery plantings reduces the cutting yield.

Cultural

Since the adults hibernate under bark, litter and forest debris, proper sanitation practices, in and around nurseries, should be adopted to either destroy the hibernating beetles directly or to expose them to winter temperatures.

Biological

The spring generation of the leaf beetle may be greatly reduced by ladybird beetles, *Coleomegills maculate*, which feed on the eggs and pupae.

Chemical

Insecticide applications should be done before the larvae enter the pupal stage to maximise the damage to predator populations.

Conclusion

The growth and productivity of forests are adversely affected by frequent outbreaks of pests and diseases. The damage to nursery stock by forest insects is often considerable at times. The most important among the nursery pests are the white grubs, cutworms and wireworms, which cause considerable damage, whereas pests like green bug and mole cricket are the minor pests. Limited research has been carried out on the different aspects of pest management in forestry under Kashmir conditions. A detailed study highlighting the major pests and the damage caused by them as well as the losses incurred needs to be carried out.

References

Cipiao, L., Bandeira, R.R. and Sitoe, S. M. 2009. Incidence of Hypsipyla sp. (Lepidoptera: Pyralidae) and its population distribution on Khayaanthoteca (Meliaceae) stands in the Manica Province, (Central Mozambique). In: Proceedings of XIII World Forestry Congress, Buenos Aires, Argentina.

Dumroese, R. K. 2012. Forest Nursery Pests (review). Native Plants Journal. 13: 257-257. FAO.2005. Global Forest Resources Assessment 2005 – India – Country Report. Forestry

Department, Forest Resources Assessment 2005, Country Report 001, 128 pp. FAO.2006. Global Forest Resources Assessment 2005 – progress towards sustainable forest management. Forestry Paper No. 147. FAO, Rome.

Lim, G.T., Kirton, L. G., Salom, S.M., Kok, L. T., Fell, R. D. and Pfeiffer, D. G. 2008. Mahogany shoot borer control in Malaysia and prospects for biocontrol using weaver ants.Journal of Tropical Forest Science, 20(3): 147–155.

Mohandas, K., Mathews, G., Nair K.S.S. and Menon, A.R.R. 1990.Pest incidence in natural forests – a study in moist deciduous and evergreen forest of India.In Pests anddiseases of forest plantations in the Asia-Pacific Region.Proceedings of the IUFROWorkshop. FAO, Regional Office for Asia and the Pacific (RAPA), Publication 1990/9,pp. 129-134.

Morris, C.L. and Hoffard, W.H. 1989. Sawflies. In: Cordell, C.E., Anderson, R.L., Hoffard, W.H., Landis, T.D., Smith, Jr., R.S., and Toko, H.V., tech. coords. Forest nursery pests. Agriculture Handbook 680. Washington, DC: USDA Forest Service: 82–83.

Singh, A.P., Bhandari, R.S. and Verma, T.D. 2005. Important insect pests of poplars in agro-forestry and strategies for their management in northwestern India. Agroforestry Systems, 63(1): 15-26.

17

Damping-off of Forest Nurseries and Its Management

T.A. Shah and K.A. Bhat

Division of Plant Pathology, Faculty of Agriculture, SKUAST-K, Wadura Jammu and Kashmir, India

Soil-borne plant pathogens cause significant losses in the nursery plantation of agricultural, horticultural and forestry plantation stands. These pathogens can survive in the soil for many years through their resistant structures. Besides, the diseases caused by them are difficult to predict. Soil factors like soil type, texture, pH, moisture, temperature, level of organic matter and nutrient content influence the activity of soil-borne pathogens and diseases. In order to survive successfully, these pathogens compete with others for nutrition, space and other requirements. Understanding the dynamics of these pathogens and other microbial populations in the soil in relation to environmental conditions is essential for the successful management of these diseases.

Among many stresses on forest plantations, the biotic and abiotic factors are of paramount importance. The National Forest Policy had set a goal that 33% of the country's geographical area should come under forest and tree cover. However, according to an estimate, forests occupy only 19.4% of the total geographical area of the country. Among biotic factors, fungi play a key role vis-à-vis disease development in forest nurseries. Therefore, disease management is an important consideration under different forest systems. The rampant or widespread use of pesticides, as in the agriculture or horticulture sector, cannot be recommended in the case of forest stands. Hence, the integration of disease management practices with insect and weed control and general nursery production practices is desirable for a sustainable plant production system. Integrated disease management involves the selection and application of a harmonious range of disease control strategies that minimise losses while maximising the plant health and number. The IDM strategy effectively prevents the epidemic progression of the disease and minimises its the impact on plant growth and productivity.

During the past several decades, massive afforestation programmes are underway in all parts of India. Millions of saplings have been planted, not only on government lands, but also on wastelands, community lands and privately owned lands. This has made the role of the forest nursery pivotal for raising quality saplings.

Nursery management is now a specialised field and intensive care of the plants is required for producing high quality planting material.

Diseases have an adverse effect on nursery stock. In general, the following precautions help to reduce the incidence of diseases in nurseries:

- The location should be an area free from known sources of disease.
- Good fertile soil with light texture and moderately acidic composition helps to prevent the occurrence of diseases.
- The nursery must be easily approachable for better supervision by trained staff.
- Experienced and trained staff must visit the nursery as often as possible so as to recognise the symptoms and signs of any incidence of disease that may have infected the nursery stock.
- Adequate drainage facilities must be ensured to prevent unnecessary dampness.

Among the various nursery diseases in India, the damping-off disease is the most widespread. It primarily affects coniferous species, though many broad-leaved species are also attacked. It is difficult to predict the loss due to this disease or estimate the area that will be affected by it. In a particular year or season, the entire nursery stock of even large nurseries may be completely destroyed, while in other years, the effect of the disease is negligible. Damping-off disease is prevalent in many forest nurseries including tropical pines.

Damping-off

Causes: A number of soil fungi may cause the disease. They are usually soil saprophytes, but under favourable conditions they tend to become pathogenic. In India, *Rhizoctonia solani* is the main cause of damping-off disease in forest nurseries. Other pathogens belong to the genus *Pythium, Fusarium* and *Phytophthora*.

Stages: Damping-off fungi may cause mortality in three phenological stages of seedlings in the nursery.

1. Pre-emergence damping-off/blight
2. Post-emergence damping-off/blight
3. Root rot

Factors Affecting Damping-off

The occurrence or intensity of this disease is influenced by a number of conditions or pre-disposing factors:

- Prevalence is more in clayey and wet soils due to anaerobic conditions on the rooting 231 zone.
- Damping-off is favoured in relatively high temperatures as the fungi develop rapidly due to the warmth. This disease develop more in summer or late sowings than in seedlings which have emerged during or before spring.
- The soil pH too has an influence on damping-off. The effect of this disease is more in alkaline or neutral soils than in acidic soils.
- Damping-off is also favoured by low intensity light.
- The severity is also favoured under the dense sowings, application of nitrogen fertilizer in high doses or addition of raw, undecomposed humus.

Management Measures

Cultural practices

These are considered as the first line of defence, by virtue of which, the initial inoculum load or density can be reduced through the selection of appropriate planting material, destruction of crop debris, elimination of living plants that carry pathogen and sometimes, crop rotation. The sowing practices such as the time and depth of sowing also influence the disease severity. Formalin drenching of the soil, with subsequent cover by polyethylene sheets, is effective is destroying pathogen propagules like sclerotia.

Plant Nutrition

Plant nutrition is an important consideration in increasing the plant vigour to withstand the onslaught of disease. A well- balanced supply of soil nutrients will result in healthy vigorous plants. However, many pathogens like biotrophs and viruses thrive well under ideal growth conditions. The excess use of nitrogen fertilisers can result in more vegetative growth and more succulence and hence, the plants become more susceptible to diseases. Potassium generally inhibits disease development and counteracts some of the disadvantages of nitrogen fertilisers. The phosphorus fertilisers influence the diseases in different ways in different plants. Calcium is also an essential nutrient for the composition of cell walls. An adequate supply of calcium makes cell walls more resistant to the penetration of facultative pathogens.

Use of Compost

The addition of compost into soils is a fundamental cultural practice, specially in organic productions. It improves the soil fertility and other physical conditions. The application of stable and good quality compost results in enhanced population of beneficial and antagonistic microbes, which inhibit the pathogen population in nursery stands. The compost is more suppressive to soil-borne diseases. Readily available carbon compounds from low quality immature composts support soil pathogens like *Pythium* and *Rhizoctonia*. The comparative saprophytic ability of pathogens and antagonists will determine the survival and population of microbes and pathogens in the soil.

Use of Antagonists

The nursery plantation losses can be reduced by the use of biocontrol agents, viz., *Trichoderma viride, T. harzianum, Pseudomonas, Bacillus subtilis* etc. The talc- based bioagent formulations are mostly used as dry seed dressings for controlling soil-borne diseases. The mode of antagonisms are mycoparasitism, antibiosis, competition for nutrients and space, induction of host resistance, siderophore formation etc. The efficacy of disease control of bioagents can be enhanced using organic substrate. The combined use of bioagents with fungicides also improves the disease control potential of bioagents. The locally isolated bioagents with high CFU are essential for suppressing the pathogens.

Use of Chemicals

Both the seed and soil be treated with chemicals for controlling the incidence of damping-off and other seed or soil-borne diseases. The judicious use of fungi toxicants like captan, zineb, mancozeb, copper oxychloride, hexaconozole or carbendazim can prove efficacious in the adequate management of such diseases. The use of these chemicals has been found to be fairly effective. These may either be mixed with the soil or broadcast and forked into the topsoil, a day before the seed is sown.

Last but not the least, the accurate and timely diagnosis of nursery disease(s) is an essential component of integrated disease management. To fulfill this purpose, regular monitoring and surveillance of the nursery stands is of paramount importance.

References

Bakshi, B. K. 1976. Forest Pathology: Principles and Practices in Forestry.Forest Research Institute. Dehradun, India. pp. 362.

Butt, T.B.; Jackson, C. and Magan, N. 2001. Fungi as Biocontrol Agents. CAB International, Surrey, Kew, England. pp. 360.

Chawla, N. and Gangopadhyay, S. 2009. Integration of organic amendments and bioagents in suppression of wilt disease caused by Fusarium oxysporum. Indian Phytopathology62: 209-216.

Engelhard, A.W. 1989. Soil-borne plant pathogens----management of disease. American Phytopathological Society, St. Paul, Minnesota, USA. pp. 217.

Gangopadhyay, S.; Gopal, R. and Godara, S. L.2009. Effect of fungicides and antagonists on wilt disease. Journal of Mycology and Plat Pathology39: 331-334.

Gngopadhyay, S. and Joshi, R. L. 2000. Efficacy of Trichoderma in controlling root rots. Indian Phytopathology Golden Jubilee Proceedings1: 325-326.

Khan, M. and Gangopadhyay, S. 2008. Efficacy of Pseudomonas fluorescens in controlling root rots. Journal of Mycology and Plant Pathology38: 580-587.

Mukhopadhyay, A. N.; Shresta, S. M. and Mukherjee, P. K. 1992. Biological seed treatments for control of soil-borne plant pathogens. FAO Plant Protection Bulletin40: 21-30.

Negi, S. S. 2002. An Introduction to Forest Pathology. International Book Distributors, DehraDun, India. Pp. 239.

Zaidi, N. W. and Singh, U. S. 2004. Mass multiplication and delivery of Trichoderma and Pseudomonas. Journal of Mycology and Plant Pathology34: 732-741.

18

Sustainable Livelihoods Through Entrepreneurship in Forest Nursery Establishment

M.A. Islam ***and G.M. Bhat***

Faculty of Forestry, Sher-e-Kashmir University of Agricultural Sciences & Technology of Kashmir, Benhama, Ganderbal, Jammu and Kashmir, India

Livelihood connotes the means, activities, entitlements and assets by which people make a living, attempt to meet their various consumption and economic necessities, cope with uncertainties and respond to new opportunities (De Haan and Zoomers, 2003).

Livelihood is a set of economic activities, involving self-employment and/or wage- employment by using one's endowments (human and material) to generate adequate resources (cash and non-cash) for meeting the requirements of self and the household.

Ideally, a livelihood should keep a person meaningfully and gainfully occupied in a sustainable manner with dignity

Livelihood, therefore, goes far beyond generating income. A livelihood is much more than employment.

Sustainable Livelihood Framework Includes

- Physical capital (housing, vehicles, agricultural machines, communication facilities, transport infrastructure, irrigation works, electricity, markets, clinics, schools, bridges etc.)
- Natural capital (forest, land, water, flora, fauna, pasture, biodiversity etc.)
- Financial capital (cash assets, remittances, savings, livestock, income levels, variability over time, access to credit, debt levels etc.)
- Human capital (education, knowledge, labour availability, household size, skills, health etc.)

- Social capital (rights or claims, friends, kin, support from trade or professional associations, families, communities, committees, businesses, voluntary organisations, political claims etc.) (Carney, 1998)

A livelihood is sustainable when it can cope with and recover from stress and shocks, maintain or enhance its capabilities, assets and entitlements, both now and in the future, while not undermining the natural resource base (Chambers and Conway, 1992).

Factors for Unsustainability in Land-based Livelihoods

- Small size of land holding, low soil fertility and land productivity
- Traditional, locally made, low cost, manually operated and unscientific tools, equipments and machineries
- Low literacy, low awareness, lack of skill and large herd size
- High incidence of poverty, large- sized families and high livelihood pressure

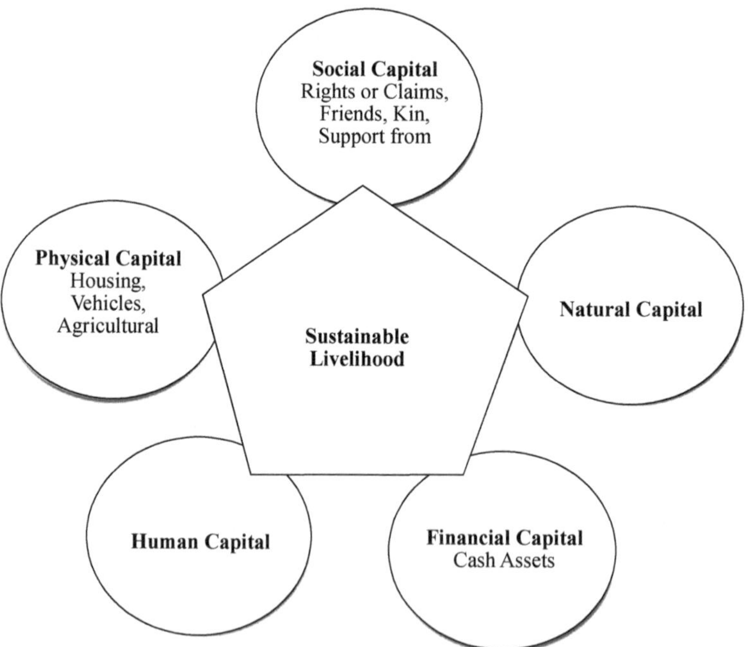

- Agriculture as main occupation, lower income and high dependence on farm income
- Limited employment and income opportunities
- Low social participation and political contact

- Unfavourable attitude towards innovations
- Limited use of information sources and extension contact
- Inadequate urban contact

Livelihood Diversification Through Entrepreneurship in Forest Nursery Establishment

Overview

Livelihood diversification is a process by which household members construct a diverse portfolio of activities and social support capabilities for survival, improvement in standards of living, income enhancement and vulnerability reduction to different livelihood shocks (Ellis, 1998). Hence, here, livelihood diversification refers to the attempts made by individuals and households to find new ways to raise income and reduce vulnerability to different livelihood shocks. Livelihood diversification can take place through both agricultural diversification, i.e., production of multiple crops or high-value crops and non-agricultural diversification, i.e., undertaking small enterprises or choosing non-agricultural sources of livelihood like casual labour or migration. The livelihood diversification through nursery raising of QPM of conifers and broad-leaved species includes various forms of wholesale and retail trade, property ownership and sales, waged employment, farming (subsistence and commercial) and the gathering and selling of wildings. The entrepreneurship in the production of QPM of forest tree species is low cost, requires low skills and is an easily implementable and highly profitable venture capable of meeting the national goal of doubling the farmer's income.

Facilities required

• Land strengthening	Equipment/ machinery Poly houses	Infrastructure
• Field store	Irrigation facility	Nursery tools

Production of QPM

The production of QPM of forest tree species *viz., Populus deltoides, Salix alba, Cedrus deodara, Cupressus torulosa, Pinus wallichiana, Thuja orientalis etc.* requires scrupulous knowledge and skill with regard to the following methods/ practices (Krishnan *et al.*, 2014).

1. Regeneration Methods of QPM

QPM of tree species can be raised by the following regeneration methods:

a) **By seed:** The tree produces seeds at intervals during good seed years and abundant seeds are collected, dried and stored for propagation.

b) **By vegetative means**:

i) **Cuttings**: Cuttings 18-25 cm long and 1-2.5 m thick having at least 4 buds are directly planted in the field in well- worked soils, keeping one bud above the ground. The soil around the cutting is thoroughly compacted.

ii) **Sets**: A set is a one- year- old cut back plant without roots or a long stem cutting which consists of one year growth. The root portion left in the nursery can again be used to produce plant/set/cuttings for the next planting season. Sets are planted in 75 cm to 1m deep pits. The length of the sets is about 3.5m or more.

iii) **Barbatelles**: Barbatelles are produced by cutting back one-year- old plants in the nursery so as to obtain a well- developed new shoot of the year in the following year. It is possible to get a one- year- old stem on a two- year- old root by this means.

iv) **Bag plants**: These can be raised in polyethene bags of 15 cm x 23 cm x 150 gauge, filled with soil mix and treated with 5% Aldrex.

v) **Stumps (root shoot cuttings):** These are prepared from one- year-old nursery plants, keeping 25 cm long root and 3cm of shoot.

vi) **Entire transplants (ETPs):** These are one or two- year- old plants raised from cuttings in the nursery.

2. Collection of Planting Material

a) Selection of plus tree

The characteristics which are kept in mind while choosing the source of planting material are:

- The trees should have long, cylindrical bole without forking.
- The trees with vigorous branching were discarded.
- Disease- free and healthy trees should be selected for preparing cuttin.

b) Seed collection & storage

Cones/ seeds are collected from October-November during seed years before they start breaking up. They are dried in the sun for a week to enable the scales to open and later, thrashed, cleaned and sieved to separate the seeds.

c) Collection of cuttings

Cuttings are collected from phenotypically superior plus trees in December-February at the time of dormancy. Cuttings obtained from young, healthy and vigorous plants perform better. The size of the cuttings should be 18-25cm containing 3-4 buds.

3. Field Preparation

a) Site Selection

Before a field is used, collect the history of the field, including previous crops grown and types of herbicides and pesticides previously applied.

Choose a site with well- drained soils, free from flooding and rocks and with a high water table.

b) Soil Types

Test soils for pH, phosphorus (P), potassium (K) and certain micronutrients. Soil pH should range from 6.0 to 6.5 for most plants and lower (5.0 to 6.0) for acid- loving plants.

Check soil survey maps for data on the soil texture of your site.

c) Slope

Avoid steep slopes to prevent erosion and low areas which are prone to cold pockets. Gentle sloping land is the best.

d) Runoff water management

- Lay out and plant fields across slopes and on contours.
- Provide grassed roadways and vegetative aisles between the rows when the topography creates erosive conditions.
- Install field border strips to reduce the movement of sediment from a field.
- Seed the areas between rows of trees with a green manure crop, such as winter rye or a more permanent crop, such as a turf grass mix, to prevent erosion.
- For information on conservation practices to prevent erosion, contact your local NRCS office.

e) Irrigation

- Plan for an adequate water supply for irrigated field nurseries. Growing nursery stock may require 1 to 2 applications of an inch of irrigation per week.
- Design an irrigation system as part of the plan for field layout and planting strategy.
- Use drip irrigation when possible. Drip irrigation conserves water and fertiliser and reduces weed competition.

4. Sowing/ Planting

a) **Direct sowing:** Seeds are sown by the broadcast method in November or in contour lines 3 m apart, in well- worked 30 cm wide terraces, continuous or broken or in patches of various sizes, viz., 1.5m x 0.5m to 1.5m x 1.5m, spaced 2m apart.

b) **Planting:** Always set cuttings with the top end up and pointing downstream at an angle of 45 to 60 degrees. Completely bury the cuttings to within an inch or two of the top bud. Always firm each cutting solidly with your foot. It is recommended that holes be made for the cuttings with a probe to avoid damaging the buds when the cuttings are inserted into the ground. A steel rod makes a good probe for this purpose.

5. Intercultural operations

- Weeding is done at 15 days interval. Irrigation is done at 3 days interval.
- The application of insecticide/ pesticide/ manure/ fertiliser is done whenever required depending on the species.
- The seedlings are pricked out twice to encourage fibrous root development.
- Casualty replacement is also necessary to maximise the production.

6. Assessment of QPM

There are two categories for the assessment of QPM of forest trees:

- **Morphological Quality**: Height and stem diameter are the two characteristics which are most commonly examined on forest seedling stock.
- **Physiological Quality**: It includes examining the frost hardiness, root growth potential, plant moisture stress, bud development etc.

7. Economics of production of QPM of forest species

The benefit-cost ratio is based on the predicted values of gross returns and the actual cost incurred during the production of QPM.

A. Cost of production of QPM of *Populus deltoides* (3775 ETPs)

A1. Variable costs (0.203 ha)

Item	Quantity	Rate (Rs.)	Cost (Rs.)
Ploughing by tractor	-	800/ hour	800.00
Levelling and bed preparation (Labour)	16 man-days	200/ man-day	3200.00
Planting material cuttings	3775	1.75/ cutting	6606.25
Irrigation (Labour)	22 man-days	200/ man-day	4400.00
Weeding	27 man-days	200/ man-day	5400.00
Miscellaneous costs	-	-	3000.00
Variable cost	-	-	23406.25
Interest on variable cost @ 8%	-	-	1872.50
Total	-	-	**25278.75**

A2. Fixed costs (0.203 ha)

Item	Quantity	Rate (Rs.)	Cost (Rs.)
Rental value of land	0.203 ha	50000/ ha	10250.00
Nursery tools			
Tungru	6	250/ Tungru	1500.00
Khurpi	8	80/ Khurpi	640.00
Spade	5	250/ Spade	1250.00
Rose Can	5	300/ Rose Can	1500.00
Secateurs	4	900/ Scateur	3600.00
Garden Rake	2	200/ Garden Rake	400.00
Auger	3	150/Auger	450.00
Rope	200 meters	2/ Meter	400.00
Measuring Tape	1	500/ Tape	500 .00
Sub-Total	-	-	10240.00
Depreciation on nursery tools @ 10%	-	-	1024.00
Fixed cost	-	-	19366.00
Interest on fixed capital @ 10%	-	-	1936.60
Total fixed cost	-	-	**21302.60**

Cost of production of 3775 ETP's = 46581.35

Total returns (@ Rs. 25/ETP) = Rs. 94375.00

Benefit/ Cost Ratio (BCR) = 94375.00/ 46581.35

= 2.03

B. Cost of Production of QPM of Cedrus Deodara and Cuppressus Torulosa (5000 seedlings)

B1. Variable costs (0.005 ha)

Item	Quantity	Rate (Rs.)	Cost (Rs.)
Bed preparation (08 beds) labour	6 man-days	200/ man-day	1200.00
Seed			
Cedrus deodara	400 gm	800/ kg	320.00
Cuppressus torulosa	200 gm	750/ kg	150.00
Transport charges	-	-	1000.00
Seed sowing (Labour)	1 man-day	200/ man-day	200.00
Polybags	5000 (23.79 kg)	165/kg (1kg =210 polybags) 1p.b. = Rs. 0.80	4000 .00
Potting media (Soil + Sand)	70 cft (Sand)	2400/90 cft (1 tractor)	1950.00
Fertiliser (DAP)	14 kg	16/ kg	224.00
Filling of polybags	23 man-days	200/ man-day	4600.00
Pricking and transplanting of seedlings	12 man-days	200/ man-day	2400 .00
Irrigation	28 man-days	200/ man-day	5600.00
Weeding	13 man-days	200/ man-day	2600 .00
Labour cost (Marketing)	6 man-days	200/ man-day	1200.00
Miscellaneous Cost	-	-	5000 .00
Variable Cost	-	-	30444.00
Interest on variable cost @5%	-	-	1522.20
Total	-	-	**31966.20**

B2. Fixed costs (0.0037 ha)

Item	Quantity	Rate (Rs.)	Cost (Rs.)
Rental value of land	0.0037 ha	50000/ ha	185 .00
Polyhouse rent	-	800/ month	19200 .00
Depreciation	-	-	-
Tungro (25/t)	6	250/ Tungru	1500.00
Khurpi (8/k)	8	80/ Khurpi	640.00
Sub-Total	-	-	19664 .00
Interest on fixed capital @ 10%	-	-	1966.40
Total fixed cost	-	-	21630.40

Cost of production of 3775 ETP's = 53596.60

Total returns (@ Rs. 20/ Seedling) = Rs. 100000.00

Benefit/ Cost Ratio (BCR) = 100000.00/ 53596.60

= 1.86

8. Marketing of QPM of Forest Species

QPM of forest species has a vast regional market and the agroforestry farmers, forest departments, industries, nursery plant sellers etc. are the major consumers.

Determinants of Entrepreneurship in Forest Nursery Establishment

General/ Personal: Burden of work, tension and challenge, leisure time, health problems, unsystematic planning and working etc.

Physical: Lack of infrastructure, equipment, land, seed/ seedling sources, skilled labour etc.

Social: Social participation, educational status, acute poverty, community/ group approach, leadership etc.

Economic: Knowledge about the marketability of the products and developed markets for the products, familiarity with alternative marketing approaches, apparent profit potential etc.

Psychological: Knowledge and skill, interest, motivation, risk- taking ability, attitude, aspiration etc.

Financial: Adoption or start-up costs, financial assistance, expense of additional management etc.

Environmental: Climatic conditions, landslides, flood, drought, insufficient water etc.

Extension and Communication: Training or expertise, availability of information about forestry, technical literatures, information sources, extension contact etc.

Technical & Managerial: Familiarity with technologies, technical assistance, availability of consultancy services and skilled managers etc.

Institutional: Demonstration sites, trained or experienced personnel in extension and communication, hierarchical relationship between the extensionagent and the client, participatory planning process etc.

Motivation- key Instrument for Entrepreneurship in Forest Nursery Establishment

A **motive** is some inner drive, impulse, intention etc. that causes a person to act in a certain way to achieve a goal he or she considers to be important at a particular time within their cultural environment (Reddy, 1993). **Motivation** is defined as the process that initiates, guides and maintains need-satisfying and goal-seeking behaviours or activities. It involves the biological, emotional, social and cognitive forces that activate behaviour (Maslow, 1971). Motivation among people can originate from specific needs, wants, desires, motives, incentives or urges, such as:

- **Need for security**: Economic, social, psychological and spiritual security.
- **Need for affection or response**: Companionship, gregariousness and social mindedness.
- Need **for recognition**: Status, prestige and achievement.
- **Need for new experiences**: Adventure, new interest, new ideas, new friends and new ways of doing things.
- **Organic needs**: Organic needs like sex, hunger and thirst are also very important for human beings.

Importance of Motivation in Extension and Development of Entrepreneurship in Forest Nursery Establishment

- Motivation is necessary for mobilising the people and extension workers.
- Knowledge of biological drives/ needs helps the extension worker to realise the problems of the people. It helps in sympathetic handling of issues.
- Knowledge of psychological and social drives helps the extension worker to formulate programmes and make effective approaches in changing their attitude.
- Knowledge of other motivating forces helps in avoiding conflicts or tensions (Dhama and Bhatnagar, 1998).

Kinds of Motivation

- **Intrinsic motivation**: It is the stimulation that drives an individual to adopt or change a behaviour for his or her own internal satisfaction or fulfillment. Intrinsic motivation is usually self-applied and springs from a direct relationship between the individual and the situation.
- **Extrinsic motivation:** It is the drive to take action that springs from outside influences instead of from one's own feelings and often involves rewards such as trophies, money, social recognition or praise. It occurs when an incentive or goal is artificially introduced into the situation to cause one to accelerate activity.

Process of Motivation

There are many stages involved in the process of motivation:

- ***Unsatisfied needs and motives:*** In this stage, unsatisfied needs can be activated by internal stimuli such as hunger and thirst. They can also be activated by external stimuli such as advertisement and window display.
- ***Tension:*** In this stage, unsatisfied needs create tension in the individual. Such tension can be physical, psychological and sociological. In this situation, people try to develop objects that will satisfy their needs.

- ***Action to satisfy needs and motives:*** In this stage, tension creates a strong internal stimulus that calls for action. The individual engages in action to satisfy his needs and motives so as to reduce tension. For this purpose, he searches for alternatives and a choice is made. The action can be hard work for earning more money.
- ***Goal accomplishment:*** In this stage, the action taken to satisfy needs and motives accomplishes goals. It can be achieved through reward and punishment. When actions are carried out as per the tensions, then people are rewarded. If not, they are punished. Ultimately, the goals are accomplished.
- ***Feedback:*** This is the last stage of motivation. Feedback provides information for the revision, improvement or modification of needs as required. Needs and motives are modified depending on how well the goal is accomplished. Drastic changes in the environment can necessitate the revision and modification of needs (Maslow, 1971).

Factors of Motivation

Many factors can motivate people to accept new ideas and practices, which may vary considerably across different societies (Ahmed, 1991). Some of these are as follows:

- ***Psychological factors:*** New experiences, greater efficiency, security of earnings or output, recognition within the community, better life for children, more leisure time etc.
- ***Social factors:*** Higher social status, greater prestige, role expectancy, sociability, hobbies etc.
- ***Economic factors:*** Secure food supply, better houses, safe drinking water, health care, clothing, consumer goods, education of children, higher levels of training, more earning power etc.

The Motivator Must have the Following Skills

- Understand what drives people
- Inspire
- Communicate
- Involve
- Challenge
- Encourage
- Set an example
- Develop and coach

- Praise
- Reward (Bahal, 2005)

Implementation, extension and management

The institutions involved include:

- State departments, such as forest, rural development, tribal welfare, livelihood promotion etc.
- NGOs, CBOs and GOs
- Universities, KVKs, research institutes etc.
- Village institutions, such as gram panchayats, communal societies, self-help groups, forest protection committees etc.
- Marketing bodies

Impacts of Forest Nursery Establishment on Livelihood

The establishment of forest nurseries can bring about significant social, financial, natural, human and physical impacts that are desirable for the society (Gangadharappa *et. al.*, 2010; Arunachalam and Arunachalam, 2012).

1. ***Social capitals:*** The social impacts of forest nursery establishment are as follows:

- *Nature of occupation:* With the establishment of forest nurseries, people stop engaging in their traditional professions like hunting, gathering forest products etc. and concentrate only on nursery raising.
- *Migration:* Forest nursery establishment facilitates increased self-employment opportunities, resulting in a gradual decrease in migration.
- *Communication exposure:* Forest nursery establishment necessitates people to come in contact with field extension functionaries, radio, newspapers etc. to gain more information.
- *Subsidiary activities:* Nursery raising of QPM of industrial/ medicinal plants diversifies subsidiary activities, such as mat- weaving, basket-making, hookah pipe- making, medicinal formulation etc.
- *Food habits:* Nursery raising of food plants helps in the production of various types of food, resulting in a variation of food habits among the people.

2. *Financial capitals:* The financial outcomes of forest nursery establishment are as follows:

- *Family income:* With forest nursery establishment, people start earning more income by selling the QPM every year.

- *Employment status:* Forest nursery establishment provides employment to the local people at their doorstep all year round.
- *Livestock possession:* Nursery operations supplement good and cheap fodder/ herbage, which in turn, increase livestock production.
- *Supplementary income:* Due to the availability of raw materials, the promotion of subsidiary occupations like mat- weaving, basket- making, medicinal formulation, sheep / goat rearing etc. contributes to increased family income.
- *Agricultural support:* Live fencing, composting, manuring, mulching, nitrogen fixation etc. in the forest nursery support the day- to- day agricultural activities.
- *Energy security:* The wastes of the forest nursery are good sources of fuel and support the household energy security.

3. ***Natural capital:*** The natural impacts of forest nursery establishment on people are as follows

- *Biomass production:* There is a significant increase in biomass production with the adoption of entrepreneurship activities of forest nursery establishment. The nursery raising of a combination of timber, fodder, fuel wood and fruit species adds organic matter to the soil.
- *Groundwater recharge:* Several studies explicitly indicate that there is a significant improvement in the ground water availability due to forest nursery establishment, as it harvests excess runoff rainwater and recharges the groundwater table.
- *Dependency on natural forests:* The forest nursery establishment reduces the dependency of the people on the natural forests for QPM.
- *Micro-climatic modification:* There is a significant difference in the atmospheric temperature and edaphic characters due to the establishment of forest nursery, as the QPM contributes to the modification of micro-climatic parameters.
- Carbon sequestration and climate change mitigation.
- Biodiversity conservation and protection of wildlife habitat.
- Pollution reduction etc.

4. ***Human capital:*** The potential human services provided by forest nursery establishment include:

- Maintenance of the local cultural heritage
- Creation of recreation opportunities

- Enhancement of the landscape
- Preservation of spirituality, values, beliefs, customary rituals, habits, totems, festivals, taboos, folklore, traditional recipes etc.

5. *Physical capital:* The physical capital comprises the capital that is created by the processes of economic production and the basic infrastructure and consumer goods needed to support livelihoods. The physical capital of entrepreneurship activities include:

- Road and transport
- Market facilities
- Housing
- Machineries and equipments
- Communication facilities
- Transport infrastructure
- Irrigation works
- Electricity etc.

Conclusions

- The entrepreneurship in nursery raising of QPM of conifers and broad leaves will lead to sustainable livelihoods by meeting all the five capital requirements (social, financial, natural, human and physical) of the local communities.
- The nursery raising of QPM provides employment opportunities for technical, skilled, semi-skilled, unskilled labour.
- The nursery raising of QPM provides opportunities for capacity building and skill upgradation for the members of the local communities.
- The nursery raising generates cash income, means for poverty alleviation, flexible working hours and opportunities for women and aged people to contribute to income generation.
- The nursery- raised QPM generates a wide scope for the establishment of fruit orchards and ornamental gardens with minimum care, cost and maintenance.
- The availability of nursery- raised QPM at the beginning of the planting season saves the time, money and effort of the farmers to raise plantations.

- The nursery- raised QPM assures the availability of genetically improved quality planting materials.
- The nursery raising of QPM are important sources of supplying seedlings for meeting the fruit, pulp and paper, fuel wood, timber and other demands of the industries.

Hence, the development of entrepreneurship among unemployed youths and other stakeholders should be encouraged. The capacity building and training in forest nursery establishments needs to be strengthened among stakeholders. The demonstration of forest nursery establishment should also be developed for the effective extension and dissemination of the intervention. Entrepreneurship in forest nursery establishments is a lucrative venture to promote the farmers' welfare, reduce agrarian distress and bring parity between the income of farmers and those working in non-agricultural professions.

References

Ahmed, M.R. 1991. Planning and designing social forestry project. In: Ahmed, M.R. (ed.), Social Forestry and Community Development, pp. 109-116. FAO, FTPP.

Arunachalam, K. and Arunachalam, A. 2012. Role of Agroforestry in human livelihoods and biodiversity management. Indian Journal of Agroforestry, 14(1): 97-100.

Bahal, R. 2005. Rural social centre for sustainable development. In: Jirli, V., De, D. and G.C. Kendadamath (ed.), Information & Communication Technology (ICT) and Sustainable Development, Ganga Kaveri Publishing House, Varanasi, India.

Carney, D. 1998. Sustainable rural livelihoods, what contribution can we make? DFID, Nottingham.

Chambers, R. and Conway, G.R. 1992. Sustainable Rural Livelihoods: Practical concepts for the 21st century. IDS, University of Sussex: Brighton, UK.

De Haan, L. and Zoomers, A. 2003. Development geography at the crossroads of livelihood andglobalization. Tijdschrift voor Economische en Sociale Geografie, 94 (3): 350–362.

Dhama, O.P. and Bhatnagar, O.P. 1998. Education and Communication for Development. Oxford & IBH Publishing Co. Pvt. Ltd., New Delhi, India.

Ellis, F. 1998. Household strategies and rural livelihood diversification. Journal of Development Studies, 35(1): 1-38.

Gangadharappa, N.R., Shivamurthy, M. and Ganesamoorthi, S. 2010. Agroforestry - a viable alternative for social, economic and ecological sustainability. My Forest, 41(2): 107-119.

Krishnan, P.R., Kalia, R.K., Tewari, J.C. and Roy, M.M. 2014. Plant Nursery Management: Principles and Practices. Central Arid Zone Research Institute (Indian Council of Agricultural Research), Jodhpur, Rajasthan.

Maslow, A.H. 1971. A theory of human motivation. Psychological Review, 50, 370 –396.

Reddy, A.A. 1993. Extension Eduation. Sree Lakshmi Press, Bapatala, Guntur, A.P.

19

Recent Approaches in Biocontrol of Plant Diseases

K.A. Bhat and T.A. Shah

Division of Plant Pathology, Faculty of Agriculture Wadura, SKUAST-Kashmir Jammu and Kashmir, India

The management of diseases and pests is a major challenge. Monoculture, excess use of fertilisers, high- yielding varieties, climate change etc have not only intensified plant diseases, but also given rise to new diseases or intensified those diseases which were earlier considered of minor importance. Although decades of research have given rise to certain promising organic and biological alternatives to chemical disease management, due to the lack of proper attention of researchers worldwide, the organic and biological disease management technologies are still not very popular.

The biological control of plant diseases, using various fungal and bacterial biocontrol agents, has emerged as a very useful strategy for the management of the diseases of vegetables, cereals, fruit and forest plants. These biocontrol agents are becoming popular because of their inherent traits, such as, they do not leave any harmful residues in food and hence, pose no hazard to human or animal health; they are safe for the environment; they do not impart resistance in the target pathogen, unlike chemicals (Monte and Llobell, 2003; Perveen and Bokhari, 2012; Reena *et al.*,2013); they are self- sustaining in nature, can establish in habitat and can give long term control and; in most of the cases, they are cheaper and pose no health hazard to the workers or handlers. However, there are certain disadvantages also associated with them, such as, difficulties in their production and formulation technology, their adaptability, less shelf life of the biocontrol formulations, delivery system etc. Keeping in view these limitations, we, at SKUAST (Kashmir), have attempted certain interventions to overcome them, many of which are successful and others are in the developmental phase.

The development of an effective biocontrol product have certain major yardsticks. The first and foremost is to search for an effective bacterial or fungal agent or organism that should strongly antagonise the disease-causing pathogen. The bioagent has to be a native one, i.e., isolated or obtained from the local environment prevailing in the region. The second and very important step in the development of a biocontrol product is the development of a complete industrial production protocol of the biocontrol product. The industrial protocol should lay emphasis on following:

1. The production technology should be as cost effective as possible, so that the product may compete in the market with other conventional (chemical) products.
2. The production technology should yield a product of superior quality. As far as an industrial biocontrol formulation is concerned, it has various steps as described under

Mass Production of Biocontrol Agents

Bacterial bioagents, mostly *Pseudomonas fluorescence* and *Bacillus subtallis* and fungal biocontrol agents like *Trichoderma gliocladium* etc. can be mass produced in liquid as well as solid state fermentation. Various methods and media have been reported for solid state fermentation, but due to certain issues, this method is not popular in the biopesticide industry. Some of these issues are as follows,

1. Solid state fermentation needs a lot of space
2. Chances of contamination of media with other aggressive saprophytes are much more hence giving qualitatively inferior product.
3. In case of fungal bio agents solid state fermentation produce more conidia and less chlamydo spores which adversely effect efficacy and shelf life of product.
4. It is labour intensive.
5. It is time consuming

Because of the above-mentioned factors, the biopesticide industry does not use solid state fermentation and uses liquid state fermentation for the mass multiplication of biocontrol agents. Hence, a technology that will ensure mass multiplication through liquid state fermentation is more desirable in the case of the industrial production of biocontrol agents (Papavizas and Lewis 1989; Lewis *et al.*, 1990).

There are two aspects of bioagent production through liquid state fermentation. The first is mass multiplication media and the second, the equipment in which mass multiplication is carried out.

Mass Multiplication Media for Fungal Biocontrol Agents

A mass multiplication medium should be cheap and easily available. For this purpose, in the case of fungal bioagents, media like yeast molasses (Papavizas *et al.*, 1984) and soy molasses (Prasad and Rangeshwaran, 2002) have been mostly used, as they are cheap and also support the efficient growth of *Trichoderma*. However, molasses are a by-product of the sugar industry and are available in areas where there is a sugar industry nearby. In areas like the valley of Kashmir or other areas where a sugar industry is not located, ingredients like molasses and yeast had to be imported.

As a result of the extensive search for a local medium which can be as efficient as the conventional molasses and yeast media, we at SKUAST-Kashmir, have achieved a breakthrough in the mass production technique by the discovery of an even cheaper mass culture medium, using one of the aquatic weed species obtained from lakes, namely, horn weed (*Ceratophyllum demersum*) and the commonly available sucrose. This media is named as **Dalweed (aquatic weed) sucrose medium for the mass multiplication of *Trichoderma***. A simple protocol has been developed, whereby, a powdered mixture of shade- dried weed and sucrose is directly added to water and after sterilisation, it is used to grow *Trichoderma* biofungicide. The medium supports the efficient growth of *Trichoderma* biofungicide as any other conventional medium and is cheaper as well. The dehydrated form of Dalweed (aquatic weed) sucrose medium for the mass multiplication of *Trichoderma* has also been developed, which can be stored for any length of time and can be supplied anywhere with ease.

Dal Weed Sucrose Medium | Molasses Yeast Medium | Dal Weed Molasses Medium

Comparative growth of Dalweed for mass multiplication of *Trichoderma*

Dehydrated Dalweed Sucrose medium for industrial mass multiplication of *Trichoderma*

Mass Multiplication Equipment (fermenter)

Fermenters/bioreactors form the back bone of any biopesticide industry involved in the commercial production of microbial biocontrol agents. A fermenter is an equipment which will provide optimum growth conditions for the mass multiplication of a particular microbe only. It is thus of prime importance and often the most costly equipment used in the biopesticide industry. The cost of the fermenter is, therefore, an issue, especially for a small entrepreneur, which often becomes a limiting factor for an investor with small capital. Moreover, besides being costly, fermenters often have a complex design and are difficult to operate. Sometimes, due to their faulty design, they yield a low quality product. At SKUAST Kashmir, we have been able to design a very cost- effective fermenter equipment, which is very cheap as compared to those currently available in the market.

The present form of a bioreactor, which was the result of a remarkable collaboration between a plant pathologist and an electrical engineer, possessed the following features. 1) A 22 litre plastic autoclavable tank (20 litre working volume). 2) Aeration by an electromagnetic air pump (18W) through fabricated metallic filters containing non-absorbent cotton. 3) A heater with a thermostat (temperature control) for self- incubation, provided with a heating indicator. 4) A temperature probe displaying real- time temperature on a monitor. 5. A photometer measuring the optical density of the fermentation medium and displaying transmittance on the monitor etc. Additionally, the fermenter had some salient features such as: a) Simple and easy to operate. 2) Cost- effective (approximately Rs. 15000/- making cost/unit of 20 litres c) Specially designed for biofertiliser and microbial biopesticide mass multiplication, but can also be validated for use in other fields of microbiology. d) The microcontroller used in the device is of 32 Kb capacity and programming space of only less then 1 Kb is used. Hence, more systems can be added in the future as per requirement. e) All accessories used are either locally available/fabricated or easily available. During the investigation pertaining to the validation testing of the equipment for mass multiplication, the results obtained during the mass multiplication of bacterial bioagents by a team of scientists indicated that the bioreactor gave a high biomass of each of the test bacteria. Besides, no contamination was observed. The reason for no contamination was the unique technique of seeding/inoculation developed and employed while inoculating the culture broth. A similar trend was also observed during the mass multiplication of fungal biocontrol agents like *Trichoderma*.

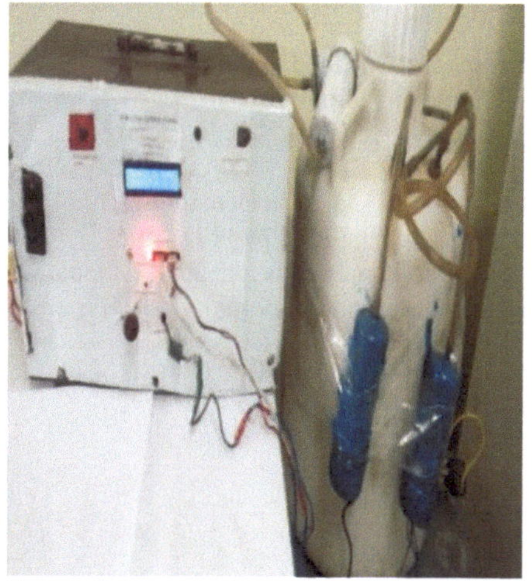

Working prototype of low cost Bioreactor

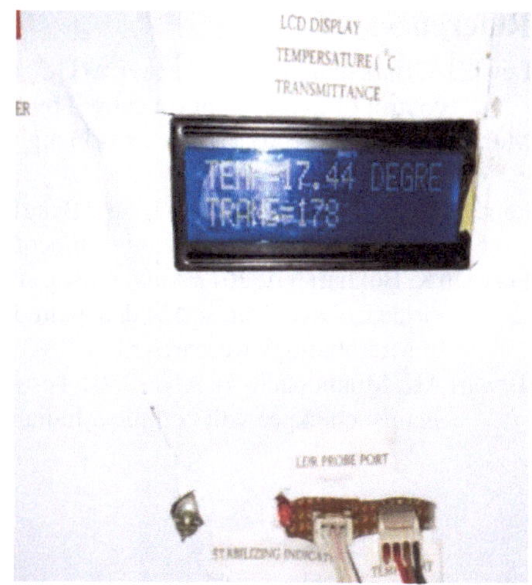

Display showing temperature and percent

Biocontrol Agents Compatible with Copper Fungicide and Triple Action Biocontrol Agents

Although biocontrol agents, if screened properly, can be very effective as protectants against plant diseases, one of their shortcomings is that they are slow in action, while disease development in various cases is fast and rapid. This flaw can be removed by developing biocontrol agents resistant to chemicals, especially those allowed in organic farming systems, like copper- based fungicide. The diversity and variability in the microbial world allows us to look for such microbes (fungal or bacterial) which are potential antagonists and at the same time, are resistant to copper. Such bioagents can be used together for the efficient management of plant diseases, especially in vegetables. At SKUAST (Kashmir), we have been able to achieve a breakthrough in obtaining copper resistant potential antagonists, which have proved useful in many ways. The synergic effect of biocontrol and copper fungicide has made disease management more viable. If used together, copper fungicides give instant control, while the associated biocontrol agent provides long- term control. Thus, this is a double action management strategy against plant diseases, especially soil- borne ones. Moving one step ahead, we have been able to obtain bioagents which are potential antagonists resistant to copper fungicide and at the same time, efficient phosphorus solubilisers. Besides, they are native. These wonderful microbes will be very helpful tools in the organic farming system. This technology of using triple action microbes is currently under validation at our centre and will be available soon for organic farming as well as integrated farming systems.

References

Lewis JA, Barksdale TH, and Papavizas GC. 1990.Greenhouse and field studies on the biological control of tomato fruit rot caused by Rhizoctonia solani. Crop Prot9 :8-14.

Monte E, Llobell A. 2003 Trichoderma in organic agriculture.VCongreso Mundial del Aguacate,, 725-733.

Papavizas GC, Dunn MT, Lewis JA and Beagle-Ristaino J. 1984.Liquid fermentation technology for experimental production of biocontrol fungi.Phytopathology 74:1171-1175.

Perveen K, Bokhari NA. 2012 Antagonistic activity of Trichoderma harzianum and Trichoderma viride isolated from soil of date palm field against Fusarium oxysporum. African Journal of Microbiology Research.; 6:3348-3353.

Tewari AK, Mukhopadhyay AN. , 2001 Testing of different formulations of Gliocladium virens against chickpea wilt complex. Indian Phytopathology; 54:67-71

20

Mycorrhizal Fungi: An Effective Nutrient Input for the Establishment of Nurseries

Zahoor Ahmad Baba

Biofertilizer Research Laboratory, Division of Basic Science & Humanities, Faculty of Agriculture Wadura, Jammu and Kashmir, India

Introduction

Soil has been advocated since the inception of nurseries as the storehouse of almost all the essential plant nutrients. Besides, it is treated as a living entity as it harbours diverse types of innumerable beneficial microorganisms, especially the beneficial genera of bacteria, fungi, actinomycetes, algae etc. In 1885, the German botanist, Albert Frank, introduced the Greek word 'mycorrhiza', which literally means 'fungus root'. These fungi form a beneficial symbiotic association with the roots of higher plants and perform the functions of root hair. This symbiotic association has been reported to promote plant growth and health by playing the role of biofertiliser and bioprotectant, respectively. Arbuscular mycorrhizal association is found in 80% of the plant species, except *Cruciferae, Chenopodiceae, Caryophyllaceae* and *Cyperaceae* (Hirrel *et al.*,1978).The association also occurs over a broad ecological range from aquatic to desert environment. The mycorrhizal symbiosis has been recognised to play a key role in nutrient cycling in the ecosystem and in protecting plants against environmental and biological stresses. In fact, many high value ornamental and edible crops enter into some form of mycorrhizal association. Most of the crop plants are mycotrophic (i.e., they have the ability to respond to AMF symbiosis). Hence, AMF may functionally benefit the productivity and /or vigour of many crops. In addition, the mycorrhizal plants have greater tolerance to toxic heavy metals, root pathogens, drought, high soil temperatures, soil salinity, adverse soil pH and transplantation shock. Because of their widespread occurrence in nature and their numerous benefits to plants, the fungi are currently attracting much attention in agricultural, horticultural and forestry research. Mycorrhizae are one among the wonderful fungi which have multidimensional beneficial effects on the soils and plants under different ecological conditions. This group of fungi is very versatile by means of improving the macro and micronutrient uptake of plants, enhancing the water relations of crops under stress conditions and protecting the roots from the adverse effects of harmful soil microbes. These fungi also contribute biomass, besides improving the physical condition of soils. Mycorrhizal fungi can be isolated and mass multiplied by easily available simple techniques. Arbuscular mycorrhizal fungi enter the root cells and cannot be seen without the aid of a microscope.

Arbuscular mycorrhizae are especially effective at transferring carbon to the soil in the form of glomalin, a sticky glue-like substance, which is estimated to provide 30 to 40% of the carbon found in soils. Forest tree species with arbuscular mycorrhizae include cedars, cypress, junipers, redwoods, maple, ash, dogwoods, sycamore, yellow-poplar and sweetgum. The agricultural crops used by forest nurseries as cover crops also form mycorrhizae. The intensive commercial seedling production for reforestation typically suppresses or delays the colonisation of seedlings by mycorrhizal fungi. Fumigation used to control pests can limit and sometimes, remove mycorrhizal fungi from the upper 15 to 30 cm (6 to 12 in) of soil, for weeks to several months. Some fungicides have also been found to suppress mycorrhizae development, especially systemic fungicides (for example, triadimefon) and fungicides applied as soil drenches (for example, azoxystrobin, iprodione). High fertilisation rates have also been known to suppress mycorrhizae development, particularly when the fertiliser is high in phosphorus (greater than 150 ppm). Despite the negative effects these common nursery treatments have on mycorrhizae, the benefits of fertilisation and pest control often outweigh the delay of mycorrhizae formation on seedlings. Ectomycorrhizal fungi are able to infest fumigated fields or containers via windblown spores produced by mushrooms and puffballs growing within or near the nurseries. These ectomycorrhizal fungi are often well-adapted to intensive nursery systems. *Thelephora terrestris* is common in most North American nurseries. In the Pacific Northwest, *Laccaria laccata* and *Inocybe lacera* are also common forest nursery colonisers. In contrast, arbuscular mycorrhizal fungi produce soil- borne spores that are unlikely to colonise container- growing media and can be slower to recolonise fumigated soils. This delay in colonisation can limit the production of species highly dependent on arbuscular mycorrhizae. Though there are different mycorrhizal associations, the most common type occurring in all ecological situations, is the vesicular arbuscular mycorrhiza (Bagyaraj, 1989; Barea and Jeffries,1995). Increased plant growth because of VAM colonisation is well- documented (Bagyaraj and Verma,1995). The increased plant growth is attributed to the enhanced uptake of diffusion limited nutrients, hormone production, biological nitrogen fixation, drought resistance and suppression of root pathogens.

Biological control can be defined as the directed, accurate management common components of ecosystems to protect plants against pathogens. Several workers have reported that VAM fungi can act as biocontrol agents for alleviating the severity of diseases caused by root pathogenic fungi, bacteria and nematodes. It is evident that an increased capacity for nutrient acquisition resulting from mycorrhizal association could help the resultant stronger plants to resist stress. However, VAM symbiosis may also improve plant health through a more specific increase in protection (improved resistance and/ or tolerance against biotic and abiotic stresses). Mycorrhizae appear to be extremely advantageous to the growth of crops in low fertility soils, which are a characteristic of poorly managed, continuously cropped agricultural lands as well as drastically disturbed landscape and mined reclamation sites. An increase in mineral uptake, as the result of mycorrhizal associations, is often reflected in increased plant survival, growth and yield as well as nutrition. The improved plant growth is attributed to increased nutrient uptake, especially phosphate, due to the exploration by the external hyphae of the soil beyond the root hair zone. The phosphate uptake is more significant in soils deficient in phosphorus. The increased growth of plants inoculated with VAM fungi is not only attributed to improved phosphate uptake, but also to the better availability of other elements like Zn, Cu, K, S, Al, Mn, Mg, Fe etc. Allen *et al.*,(1982) illustrated that VAM directly affects the levels of plant hormones like cytokinins and gibberellins substances. Plants colonised by VAM fungi can tolerate a wide range of soil water regimes and also improve water relations of many plants.

Improved Plant Nutrition

i. ***Phosphorus:*** Phosphorus is one of the most important nutrients for plant growth. It is one of the least available and mobile plant nutrients in the soil (Takahashi and Anwar, 2007). Many soils have high reserves of total phosphorus. However, only 0.1% of it is available to plants (Zou *et al.*,1992). At present, 5, 49.3, 48.8 and 1.9% of Indian soils fall under adequate, low, medium and high categories of available phosphorus status respectively. The soils of Kashmir fall under the low category of available phosphorus (Pattanayak and Tarafdar, 2009). Inorganic fraction is an important form of phosphorus in soils. It is generally categorised into insoluble and readily soluble categories. The insoluble fraction is neither available to growing plants, nor to microorganisms and constitutes 94 to 99% of the total soil phosphorus. This fraction is mostly attached to Fe and Al in acid soils and to calcium in slightly acidic to alkaline soils. Inorganic phosphates, when applied to soil, are transformed into various reaction products, mainly remaining in the sparingly soluble orthophosphates of Al, Fe and Ca. Therefore, expensive phosphatic fertilisers have to be applied to the agricultural fields to maximise production. However, the soluble phosphorus in these fertilisers is easily and rapidly precipitated to insoluble forms with cation such as Ca^{2+}, Fe^{3+}, Al^{3+} or Zn^{2+} or adsorbed to calcium carbonate, aluminium oxide, iron oxide and aluminium silicate. Therefore, the apparent recovery of applied phosphorus in soils is very less, i.e., 15 to 20%. This transformation decreases the efficiency with which soluble phosphorus can be taken up by the plants and decreases the effectiveness of the fertiliser, resulting in the application of increasing amounts of phosphatic fertilisers to the agricultural fields. This unmanaged use of phosphatic fertilisers has increased agricultural costs and instigated a variety of environmental and health hazards as these contain potentially toxic heavy metals (Pb, Cd, As etc).Their excessive use has rendered the fertile soils sick by disturbing the soil microbial biodiversity.

Therefore, the use of VAM in organic agriculture for enhancing host plant phosphorus nutrition is an economically and environmentally promising strategy.

- The external mycelium of mycorrhizal fungi can take up P in the form of trehelose phosphate more effectively than roots at low concentrations.
- The external mycelium can proliferate far beyond the rhizosphere and increases the soil volume, which is exploited for phosphorus uptake. The hyphal transport of Phosphorus has been estimated to be 20-90 % and likely to fulfill the entire requirement of fertiliser phosphorus.
- The rapid absorption of the soluble form of phosphorus by the external hyphae leads to a shift in the equilibrium towards the release of bound phosphorus from soil reserves (Smith and Read,1997).
- The mycorrhizal roots of onion increased the acid phosphatase activity by 20 to 30 times in comparison to non- mycorrhizal roots that catalyse the hydrolysis of complex insoluble phosphorus compounds in the soil and increase the soluble form of phosphorus.

These mechanisms aid in the uptake of phosphorus by the host plants and help in reducing the dependence on inorganic phosphatic fertilisers. Thus, mycorrhiza plays a pivotal role in the solubilisation, mobilisation and uptake of phosphorus and can be exploited in organic farming up to the fullest possible extent

ii. ***Nitrogen:*** Among all the essential nutrients, nitrogen has the distinction of being called the 'Kingpin' nutrient. Its use is indispensible in low as well as conventional production systems. The available nitrogen status in agricultural soils is subjected to various losses through processes like leaching and volatilisation. Under such conditions, mycorrhizal fungi play a significant role in improving the nitrogen nutrition of plants through acquisition and assimilation mechanisms.

- The external fungal mycelium plays an important role in direct nitrogen acquisition and transport to the root cells, thereby contributing to plant nutrition. Studies by Fray and Schuepp (1993) have revealed that the extraradical mycelium in mycorrhizal fungi can derive ^{15}N from the soil. Subramanian and Charest (1999), in a box compartmental experiment, have shown that the amount of nitrate (NO_3^-) ions being transported by the external hyphae was about 30-35% under water deficient conditions.
- The mycorrhizal colonisation of roots has increased the activities of nitrogen assimilatory enzymes, such as nitrogen reductase (NR), glutamin synthetase (GS) and glutamate synthase (GOGAT) in drought-stressed maize roots (Subramanian and Charest ,1998;1999).
- Under soil conditions where less mobile ammonium ions are dominant, the role of mycorrhizal symbiotic association becomes more important.
- Mycorrhizal fungi enter into tripartite association (Soybean- Rhizobium-Glomus) and thereby aid in the transfer of nitrogen fixed by Rhizobium to the non- leguminous neighbouring plants.

These evidences suggest that mycorrhizal fungi can successfully be exploited for improvement in the nitrogen nutrition of crop plants under organic farming.

iii. ***Micronutrient nutrition:*** The external hyphae explore the soil beyond the root hair zone and thereby increase plant growth by enhancing the uptake of diffusion limited nutrients. Mycorrhizal hyphae develop intensively inside the roots and within the soil, forming extensive extra radicals, which help the plant in exploiting mineral nutrients and water in the soil. In plants, particularly those with week/restricted root system, hyphal connections act as a bridge between the roots and nutrient sites in the soil and facilitate the efficient uptake of immobile nutrients by the host plant. Depending upon the host plant, colonisation by mycorrhizal fungi can increase the uptake of micronutrients, especially Zn, Cu, Mn and Fe (Rupam *et al.*,2008). Among the essential nutrients required by crops, zinc is considered the most critical micronutrient. Its deficiency can cause yield reduction to the tune of 10 to 50%, depending on the severity and stage of occurrence. The magnitude of Zn deficiency is high in almost every type of soil and the major portion of added Zn is fixed.

Further, the imbalanced use of fertilisers and non-addition of organic manures are believed to aggravate the situation. In some cases, zinc deficiency in soil reduces grain yield up to 80%, along with a reduction in grain Zn content and other nutritional qualities. The high dependence on cereal- based diets with low levels of Zn brings out malnutrition in human beings and globally, over two billion people are affected by Zn deficiency. Improving food grain production with nutritionally rich grain quality is the need of the hour to sustain grain production and ensure nutritional security. Despite the fact that the importance of Zn nutrition is well known, it is very difficult to ameliorate Zn deficiency in crops due to the extremely low use efficiency of zinc (less than 1%) in crops and the remaining 99% get fixed in the soil. Indeed, the majority of arable lands have high total Zn, but the bioavailability is too low, suggesting that there is a need to adopt strategies to transform the unavailable form of zinc to available forms. One of the biological means to mitigate Zn deficiency is by exploiting naturally occurring mycorrhizal symbiosis. Arbuscular mycorrhizal (AM) fungi immobile micronutrients such as Zn and Cu.

- Mycorrhizal fungi lower the pH around the rhizosphere that helps in the release of Zn from the fixed pool.
- The external mycelium of mycorrhizal fungi is very explorative and transports Zn far from the root zone, to the tune of 40%, thus contributing to the host plant nutrition.
- The rhizosphere of mycorrhizal roots is biochemically active in terms of soil enzymes and releases a specific glomalin protein that serves as an adsorptive site for Zn, which in turn, is made available to the host plant.
- Interestingly, mycorrhizal fungi are able to extract Zn from the tightly bound residual forms of Zn and contribute it for the organic bound and water soluble forms of Zn. As the result of these mechanisms and processes, mycorrhizal plants are more efficient in utilising Zn from the soil and help to produce higher grain yield by 10-15%. Thus, mycorrhizal inoculation is one of the potential strategies to improve Zn use efficiency by crops, besides enhancing the yield and quality of grains.

iv. ***Plant protection:*** Mycorrhizal fungi have been well- documented as biocontrol agents and the general conclusion is that they can reduce or even suppress the damage caused by soil- borne pathogens. AMF- colonised plants have shown a significant degree of bioprotection against various pathogens, like *Fusarium, pythopthora, Aphanomyces, Verticillium*

(Elsen *et al.*, 2001 and Azcon and Barea,1996) and nematodes causing root rot, lesions, wilt and galls respectively (Guillemin *et al.*,1994). Several genes and corresponding protein products involved in plant defence responses have been extensively studied in AMF symbiosis and have been shown to be spatially and temporally expressed (Harrier and Watson, 2004). These include callose deposition, phytoalexins, β-1-3 glucanases, chitinases and PR pathogenesis- related proteins (Guillon *et al.*, 2002). Cordier *et al.*,(1996) showed that the pre-inoculation of tomato with an AM fungus, subsequently challenged by *Pythopthora parasitica,* resulted in less root damage. In this study, the researchers used immune gold labelling technique to show that the number of hyphae of the pathogen was greatly reduced in mycorrhizal roots and mycorrhizal root tissues infected by the pathogen. The AMF was able to confer bioprotection against *Phytophthora parasitica* via localised and induced systemic resistance in mycorrhizal and non- mycorrhizal roots respectively.

v. ***Alleviation of environmental stresses:*** Mycorrhization with arbuscular mycorrhizae enables plants to tolerate a wide range of environmental stresses, such as drought, toxic metals, saline soil, root pathogens, high soil temperature and adverse pH (Caldwell and Virginia,1989). A well-developed mycorrhizal symbiosis may enhance the survival of plants in polluted areas by improving water relations, nutrient acquisition, pathogenic resistance, amelioration of soil structure, phytohormone production and contribution to soil aggregation, thus improving the success of all kinds of bioremediation techniques, such as decreased caesium uptake by AMF treated plants. It can also be used effectively in the establishment of plant cover on radio nuclide contaminated soils, thereby reducing environmental risks. Mycorrhization can also be used for attenuation of deleterious soil conditions. They also have the potential to monitor site toxicity or the efficiency of restoration techniques (Weissenhorn *et al.*,1993). Therefore, mycorrhizal fungi enable plants to cope with abiotic stresses by alleviating mineral deficiencies, overcoming the detrimental effects of salinity, improving drought tolerance, enhancing tolerance to pollution and improving the adaptation of sterile micropropagated plantlets to cope up with sudden stress situations arising due to changes in the environmental conditions, encountered as a result of their shift from invitro to invivo conditions (Barea *et al.*,1993).

Mycorrhizae protect the plants from the adverse impact of heavy metals by the following mechanisms:

A. Biosorption by Mycorrhizal Fungi

- Adsorption: Fungal wall (chitin) binds the metals.
- Complexation: Organic acids produced by mycorrhizae form complex? with heavy metals.
- Precipitation: Formation of intracellular heavy metal phosphates.

B. Detoxification Mechanism

- Avoidance: Sometimes, the mycorrhizal mycelium avoids the absorption of metal ions.
- Solubilisation: Dilution of metals.

Arbuscular Mycorrhizal Inoculation	60-100 spores/100 g
Optimum spore count	Soil
Rate of Inoculation: Vegetables	100 g/m^2 nursery
Fruit trees and apple	100-200 g/tree
Nursery plants	50 g/pot
Other crops	10% of seed rate

Methods of Inoculum Production

The threshold point related to the use of AM fungi as plant growth promoters is the development of suitable techniques for the production of large quantities of pure pathogen free inoculum with high infectivity potential. Some of the commonly used methods for the mass production of AM spores are listed below:

a) *Soil based inocula*

- *Pot culture:* It is the most widely used standard and conventional method of maintaining AM fungal cultures around the world .In this method, AMF spores are inoculated to the roots of a trap plant raised on sterilised soil. Though the usual substrate used in pot culture is sterilised sand-soil mixture, sometimes, inorganic inert materials like peat, perlite and vermiculite can also be used as substrate (Abdul Khaliq *et al.*,2001). The trap plants commonly used for pot culture are *Sorghum halepense, Paspalum notatum, Panicum maxicum, Cenchrus ciliaris, Zea mays, Trifolium subterraneum* and *Allium cepa* (Chellapan *et al.*,2001). The inoculum so produced consists of a mixture of soil, spores, hyphal segments and infected root pieces and generally takes around 3-4 months.
- *Inoculum- rich soil pellets:* A technique of AMF inoculum production, in which soil pellets are enriched with the AMF inoculum, was introduced by Hall and Kelson (1981). The pellets had an average dry weight of 1.55 g and measured 12mm × 12 mm × 6 mm. These dry pellets can be

glued to the seeds by gum arabica and can easily be broadcasted and spread like other fertilisers during seed sowing or transplantation.

b) *Soil-free inocula*

- *Aeroponic culture:* Although soil- based pot cultures are the most widely used method for AMF inoculum production, nowadays, the focus is shifting towards alternative soilless cultures for the mass production of clean and pure AMF propagules, for physiological and genetic studies for invitro mycorrhization (Mohammad *et al.*, 2000). In aeroponic cultures, pure and viable spores of a selected fungus are used to inoculate the cultured plants, which are later transferred to a controlled aeroponic chamber (Singh and Tilak, 2001), where the nutrient solution is provided in the form of a mist. The lack of physical substrate ensures extensive root growth, colonisation and sporulation of the fungus and makes it an ideal system for obtaining sufficient amounts of clean AMF propagules (Abdul Khaliq *et al.*, 2001).
- *Root organ culture:* The main obstacle in the study of AMF and AMF symbiosis is the obligate biotrophic and hypogeal nature of the endophyte. Several attempts have been made in the past to overcome these hurdles through the use of invitro root organ culture, because of its potential for research and inocula production. Agrobacterium rhizogenes is a gram- negative soil- inhabiting bacteria, which produces a condition called 'hairy roots', as a result of the modified hormonal balance of the tissues that makes them vigorous and allows them to grow rapidly on artificial media (Abdul Khaliq *et al.*, 2001). Once the hairy roots are ready, spores are collected either from the field or pot cultures by the wet sieving and decanting method (Gerdemann and Nicolson, 1963). Generally, two types of fungal inocula are used for initiating monoxenic cultures: extraradical spores or mycorrhizal root fragments and isolated vesicles of the fungus. In addition to the spores and root fragments, sporocarps of *Glomus mosseae* have also been used by Budi *et al.* (1999) to establish invitro cultures. After isolating the fungus from the soil, spores are surface sterilised using a suitable surfactant solution. Generally, (tween 20 and s?) solution containing a strong oxidising agent, such as chloramine T, is used for the sterilisation of AMF spores (Fortin *et al.*, 2002). The spores are then subsequently rinsed thoroughly in streptomycin-gentamycin antibiotic solution (Becard and Piche, 1992). All the steps starting from spore isolation to rinsing should be done on ice, to maintain spore dormancy. The rinsed spores should be stored at 4°C in distilled water or water agar or on 0.1 % $MgSO .7H_2O$ solidified with gellan gum, if not used immediately (Fortin *et al.*, 2002).

The final step in raising a successful invitro culture is the selection of the appropriate culture medium for dual cultivation of the partners, the host root and the AMF. The nutrient media should be carefully selected to allow the growth of the host as well as the fungus during the dual culture establishment. The root needs rich nutrient medium for its growth and the AMF normally require a relatively poor nutrient medium (Abdul Khaliq *et al.*,2001). Generally, Murashige and Skoog's (1962) and White's medium are used for establishing the dual culture of the host root and the AMF symbionts.

- *Nutrient Film Technique:* The NFT is another technique of soilless inoculum production, pioneered by Cooper (1975). In NFT, the plant roots are provided with a shallow layer of rapidly flowing nutrient solution. As a result of it, root mats are formed and the upper layer above the liquid retains a film of moisture around them. The pre- inoculated seedlings are planted into the NFT unit. The inoculum produced by this method is ideal for the production of easily harvestable solid mats of roots with more concentrated and less bulky form of inoculum than that produced by plants grown in soil- based or other solid media (Abdul Khaliq *et al.*,2001).
- *Polymer-based inoculum:* Encapsulation or entrapment of AMF in polymer materials is frequently used as a powerful means of immobilisation. It includes the encasement of AMF spores, vesicles or mycorrhizal roots within a porous structure formed 'in situ' around the biological material. In polymer- based inoculum preparation, the AMF are generally mixed with a compound which is then gelled to form a porous matrix, under conditions sufficiently mild, so as not to effect the viability of the biological material. Around 1350 combinations of natural, semi- synthetic and synthetic polymers exist for the entrapment of AMF (Vassilev *et al.*, 2005). However, the majority of techniques involving 'in situ' entrapment make use of natural polysaccharide gels including kappa carrageenan, agar and alginates. Calcium alginate is the most widely used carrier of choice for the encapsulation of AMF. In some cases, the spores of AMF can be introduced directly in synthetic seeds, which can germinate under suitable conditions and can become complete plantlets.

It can be concluded here that AM fungi can be used quite successfully as a nutrient input for the successful early establishment of a forestry nursery.

References

Abdul Khaliq., Gupta, M.L. and Alam, A (2001). Biotechnological approaches for mass production of production of arbuscular mycorrhizal fungi: Current scnerio and future strategies. In: Mukerji, K.G., Manoharachary, C., Chamola, B.P.(eds), Techniques in mycorrhizal studies. Kluwer Academic publishers,The Netherlands,pp.299-312.

Allen,M.F.,Moor,T.S.J.and Christensen,M (1982). Phytohormone change in Bouteloua gracilis infected by vesicular arbuscular mycorrhiza II: altered levels of gibberlin- like substances and abscisic acid in the host plant.Canadian Journal of Botany,60:468-471.

Azcon,A.C and Barea,J.M (1996).Arbuscular mycorrhizae and biological control of soil borne plant pathogens;an over view of the mechanism involved.Mycorrhizae.6;457-464.

Bagyaraj,D.J.1989.Mycorrhizae.In: Tropical rain forest ecosystems, Elsevier science Publishers, Amsterdam,pp 537-546.

Bagyaraj,D.J.and Verma,V(1995).Interaction between arbuscular fungi and plants:their importance in sustainable agriculture and in arid and semi arid tropics.In:advances in microbial ecology, Academic Press, London, pp 119- 142.

Barea, J.M., Azcon, R. and Azcon,A.C(1993). Mycorrhiza and crops.In: Tommerup, I.(ed), Advances in Plant Pathology, Vol.9, mycorrhiza:a synthesis.Academic press London, pp. 167-189.

Becard,G.and piche,Y (1992).Establishment of Vesicular arbuscular mycorrhiza in root organ culture;review and proposed methdology. In: Norrisi etal (eds), Mthods in Microbiology, vol.24. Academic press, London, pp. 89-108.

Budi,S.W.,Blal,B. and Gianinazzi,S (1999).Surface sterilization of Glomus mosseae sporocarps for studying endomycorrhization in vitri. Mycorrhiza, 9:65-68.

Caldwell,M.M.and.Virginia,R.A(1989). Root systems in: pearcy, R.W., Ehleringer, J.A., Mooney, H.A, Rundel, P.W. (Eds), plant physiological ecology-field methods and instrumentation. Chapman and Hall, London, pp. 367-398.

Chellapan,P., Chrasty, S.A.A. and Mahadevan,A (2001).Multiplication of arbuscular mycorrhizal fungi on roots.In: Mukerji, K.G., Manoharachary, C., Chamola, B.P.(eds),Techniques in mycorrhizal studies.Kluwer Academic publishers,The Netherlands,pp.285-297.

Chuck,S (2008).Screening evaluation of heavy metals in inorganic fertilizers. Minnesota Department of Health ,625-Nobert Street North, St. Paul Minnesota,55164,USA.pp 26. Cooper,A.J (1975).crop production in there circulating nutrient solutions.Scientia Horticulturae,3:251-258.

Cordire,C.,Gianinnzi,P. and Gianinnzi,S (1996).Colonization patterns of root tissues by phytophthora nicotiana var.parasitica related to reduced disease in mycorrhizal tomato. Plant and Soil,185:223-232.

Elsen,A.,Declerck,S.,D,Waele (2001).Efect of Glomus intraradicies on the reproduction of burrowing nematodes (Rhadopholus similis) in dixenic culture.Mycorrhiza,11:49-51.

Fortin,J.A.,Becard,G.,Declerck,S.,Dalpe,Y.,Arnaud,S.M.,Coughlan,A.P.andPiche,Y(2002) . Arbuscular mycorrhiza on root organ culures.Canadian Journal of botany, 80:1-20.

Fray,B.and Schuepp,H (1993).Acquisition of N by external hyphae of AM fungi associated with maize.New Phytol.124:221-230.

Gerdemann,J.V. and Nicolson, T.H (1963). Spores of mycorrhizal Endogone species extracted from soil by wet sieving and decanting. Trans Britanica Mycological Society.,46:235-244.

Guillemin, J.P., Gianinazzi, P., Marchal, J (1994). Contribution of mycorrhizas to biological protection of micropropagated pine apple (Ananas comosus (L) Merr) against phthopthora cinnamomi Rads.Agric.Sci.Finl.3:241-251.

Guillon,C.,Arnod,S.T.M.,Hamel,C. and Jabaji,H.S.H (2002).Differential and systemic alteration of defence related gene transcript levels in mycorrhizal bean plants with Rhizoctonia solani. Canadian Journal of Botany,80:305-315.

Hall,I.R.and Kelson,A (1981).An improved technique for the production of endomycorrhizal infested soil pellets.Newzealand journal of Agricultural Research,24:221-222.

Harrier,L.A. and Watson,C.A (2004).The potential role of arbuscular mycorrhizal fungi in the bio protection of plants against soil borne pathogens in organic and /or sustainable farming systems.Pest management science.60:149-157.

Hirrel MC, Mehravaran H and Gerdemann JW (1978). Vesicular mycorrhiza in theChenopodiaceae and Cruciferae: do they occur? Can. J.Bot., 56:2813-2817.

Mohammad,A.,Khan,A.G. and kueck,C(2000).Improved aeroponic culture of inocula of arbuscular mycorrhizal fungi.Mycorrhiza,9:337-339.

Morton, J. B (1985). Variation in mycorrhizal and spore morphology of Glomus occultum and Glomus diaphanum as influenced by plant host and soil environment. Mycologia 77: 192- 204.

Murashige,T.and Skoog,F(1962).a revised medium for rapid growth and bioassays with tobacco tissue cultures.Plant physiologia,15:473-497.

Pattanayak, S.K., P.Sureshkumar, and J.C.Tarafdar(2009).New vista in phosphorus research. Journal of Indian Society of Soil Science, 57(4):536-545.

Phillips. J.M. and Hayman, D.S (1970). Improved procedures for clearing and staining parasitic and vesicular arbuscular mycorrhizal fungi for rapid assessment of infection. Trans. Brit. Mycol. Soc., 13: 31-32.

Rupam,K.,Deepika,S and Bhatnagar,A.K (2008).Arbuscular mycorrhizae in micropropagation system and their applications. Scientia Horticulturae,116:227-239

Scholten,L.C.and Timmermans,CWM(1992).Natural radioactivity in phosphate fertilizers. Nutrient cycling in agroecosystems.43:103-107.

Singh,G.and Tilak,K.U.B.R (2001).Techniques of AM fungus inoculum production.In: Mukerji, K.G., Manoharachary,C.,Chamola,B.P.(eds),Techniques in mycorrhizal studies. Kluwer Academic publishers,The Netherlands,pp.273-283.

Smith, S.E. and D.J. Read (1997). Vesicular-arbuscular mycorrhizas in agriculture and horticulture. Chapter 16 In Mycorrhizal Symbiosis. Second edition. Smith, S.E. and D.J. Read (eds.).pp. 453-69. Academic Press, London, UK.

Subramanian, K.S. and Charest, C. (1998) Arbuscular mycorrhizae and nitrogen assimilation in maize after drought and recovery.Physiologia Plantarum 102: 285-296.

Subramanian, K.S. and Charest, C. (1999) Acquisition of external hyphae of an arbuscular mycorrhizal fungus (Glomus intraradices Schenck & Smith) and its impact on physiological responses in maize (Zea mays L.) under drought-stressed and well watered conditions. Mycorrhiza 9: 69-75.

Takahashi,S.and Anwar,M.R(2007).Wheat grain yield ,phosphorus uptake and soil phosphorus fraction after 23 years of annual fertilizer application to an Andisol. Field crops Research.101,160-171.

Vassilev, N.,Nikolaeva,I.and Vassileva,M(2005).Polymer based preparation of soil inoculants: applications to arbuscular mycorrhizal fungi.Review of Environmental Sciences and Biotecnology.4:235-243.

Weissenhorn,I.,leyval,C. and Berthelin,J (1993).Cd-tolerant arbuscular mycorrhizal (AM) fungi from heavy metal pollutedsoils.Plant and Soil.157;247-256.

Zou,X.,Binkley,D. and Doxtader,K.G (1992).A new method for estimating gross phosphorus mineralization rate in soils. Plant and Soil,147,243-250.

21

Application of Statistical Tools and Techniques in Forest Data Sets

T.A. Raja

Sr. Associate Professor (Statistics) FV Sc & AH, Shuhama, SKUAST-K, Jammu and Kashmir, India

Introduction

Data Analysis is an approach to analyse data sets to summarise their main characteristics and extract valid information. The importance of hypothesis is generally recognised more in the studies which aim to make predictions about some outcome. As it is not possible to work on the whole population, each group is represented only by a sample of observations and if another sample is drawn, the numerical values would change. This variation between different samples drawn from the same population cannot be eliminated or reduced fully. One is supposed to draw inferences in the presence of the sampling fluctuations. Statistical tools and techniques are applied to check whether the observed differences are real significant differences or non- significant ones.

Statistical hypothesis: A statistical hypothesis is some assumption or statement about a population, which may or may not be true or equivalently about the probability distribution characterising a population, which we want to test on the basis of evidence from a random sample. If the hypothesis completely specifies the population, then it is known as 'simple hypothesis' , otherwise, it is known as 'composite hypothesis.

Tests of Statistical Hypothesis

- It is a two action decision problem after the experimental sample values have been obtained, the two actions being the acceptance or rejection of the hypothesis under consideration.

Null hypothesis (H_0)

A null hypothesis is a hypothesis of no difference. A null hypothesis states that there is no significant difference between the groups under study e.g. Let us suppose that two different varieties are to be compared for their timber value.

To formulate a statistical hypothesis, let 'x' be a variable which denotes the timber value of variety A and 'y' be the variable which denotes the timber value of variety B. Here, the null hypothesis would be that there is no difference between the effects of the timber value of the two varieties. i.e.,

H_0: μ_x (mean by A) = μ_y (mean by B)

Alternative Hypothesis(H_1)

- This hypothesis states that there is a significant difference between the groups under study. In the above example, alternative hypothesis could be

  H_1 :- $\mu_x \neq \mu_y$

Levels of Significance (∝)

- There is a chance that a hypothesis may go otherwise. It is also called the size of the critical region. Let us suppose that we accept the null hypothesis. However, since we work on the basis of samples, it may not be 100% correct and may be correct up to 95%. Therefore, there is 5% possibility that the statement may go otherwise. This 95% region will be called the acceptance region and the 5% region is statistically called the level of significance and is denoted by 0.05

Types of errors

There are two types of errors:

Type I error :The error of rejecting null hypothesis (accepting H_1 hypothesis) when H_0 is true

Type II error :The error of accepting null hypothesis when H_0 is false

***Error**

		Actual state of nature H_0 is true	**H_0 is false**
false	Reject H_0	Type I Error	Correct
Decision	Accept H_0	Correct	Type II
Error			

Steps involved in testing of hypothesis

- The major steps involved in the solution of a 'testing of hypothesis' problem may be outlined as follows:

1. **Step 1- Set up a null and alternative hypothesis**: To test a hypothesis, we must first state the null hypothesis H_0 and alternative hypothesis H .

2. **Step 2- Selecting the level of significance**: The next step in hypothesis testing is the selection of '∝', the level of significance. It establishes a criterion for the rejection or acceptance of null hypothesis.
3. **Step 3- Determining the test distribution to be used**: The test statistics is the value used to determine whether the null hypothesis should be rejected or accepted. The choice of the appropriate statistical test is based on the use of the appropriate sampling distribution (the normal or any other distribution such as t-distribution).
4. **Step 4-Defining the rejection or critical region**: After stating the null and alternate hypothesis and after selecting the level of significance and the type of test statistics to be used, it is now possible to define the rejection region or critical region of the sampling distribution.
5. **Step 5- Performing the statistical test**: We are now in a position to use our sample data and perform the appropriate statistical test, like (Z,t,F, χ^2) followed by ANOVA.
6. **Step 6- Drawing statistical conclusions** : After performing the statistical test, it is possible to determine whether the sample information agrees reasonably well with the hypothesis and to make an inference about the population

Test of Significance for Large Samples (Z test)

$$z = \frac{\bar{x} - \mu}{\sigma/\sqrt{n}}$$

$$z = \frac{\overline{x1} - \overline{x2}}{\sqrt{\frac{\sigma 1^2}{n1} + \frac{\sigma 2^2}{n2}}}$$

When Sample population is large

When variance(S.D) of the parent population is Known

Test of Significance for small Samples (t test)

$$t = \frac{x - \mu_o}{S/\sqrt{n}}$$

$$t = \frac{\overline{x1} - \overline{x2}}{\sqrt{\frac{S1^2}{n1} + \frac{S2^2}{n^2}}}$$

When parent population is exact or finite

When variance (S.D) of the population is unknown

Test of Significance for Variance ratio test (F test):

$$F = \frac{\sigma 1^2}{\sigma 2^2}$$

$$F = \frac{S1^2}{S2^2}$$

Test of Significance based on Chi-Square

$$x^2 = \sum_{i=1}^{n} \frac{[Oi - Ei]^2}{Ei}$$

If the calculated value of the test statistics is greater than(>) the tabulated value at 5% or 1% level of significance, we reject our null hypothesis of non-significant difference. There is, statistically, a significant difference among the treatments.

Example1: A sample of 900 trees has a mean girth of 3.4 feet and standard deviation of 2.61. Is the sample drawn from a large population of mean 3.25 feet.

Solution

Ho:- The sample has been drawn from the population with mean

μ=3.25

H1:- μ # 3.25

Here X⁻ =3.4, n=900, μ=3.25, σ=2.61

$$z = \frac{\overline{x} - \mu}{\sigma/\sqrt{n}}$$

$$z = \frac{3.40 - 3.25}{2.61/\sqrt{900}} = 1.73$$

Since tab Z=1.96 at 5% l.s, we accept our null hypothesis.

The sample has been drawn from the same population.

Example 2: The mean weekly sales of saplings were 146. After an advertising campaign, the mean weekly sales in 22 shops for a typical week increased to

157 and showed a (standard deviation) of 7.2. Was the advertising campaign successful?

Solution:

Ho:-There was no significant impact of advertising.

H1:-There was a significant impact of advertising.

Here $\bar{X}$ =157, s=7.2, n=22, tab t at 21 d.f, 5%l.s=2.07, at 1%=2.83

μ=146

$$t = \frac{X - \mu_o}{S/\sqrt{n}}$$

t= 7.19

Since the calculated value of t is greater than the tabulated value, we reject our null hypothesis. The advertising campaign had a significant impact on sales.

Example 3:-A group of 5 seedlings of *Pinus kesiye* were inoculated with mycorrhiza and the height growth was 42,39,48,60,and 41 kg, whereas in a second group of 7 seedlings (without inoculation), it was 38,42,56,64,68,69 and 62kg. Can it be concluded that inoculation increased the height significantly.

Solution

Ho: $\bar{X1} = \bar{x2}$

H1: $\bar{X1} \# \bar{x2}$

$\bar{x1}$ = 46, $\bar{x2}$ =57,

$$S^2 = \frac{1}{n1 + n2 - 2}\left[\sum_{i=1}^{n1}(x1 - \overline{x1})^2 + \sum_{i=1}^{n2}(x2 - x\bar{2})2\right]$$

$$t = \frac{\overline{x1} - \overline{x2}}{\sqrt{\frac{S1^2}{n1} + \frac{S2^2}{n^2}}}$$

S^2 = 121.60

t=1.70

Tab t at 5% l.s at 10 d.f = 1.81,. Hence, we accept our null hypothesis.

There is no significant difference

Paired t-test for difference of means:

Example 4:- The two series are related series and the observations are paired.

The first series is a soil core sample at one depth and the second series is one at another depth in a natural forest. The observations are paired. Paired t-test for difference of means is used.

- Depth 1:- 130,125, 120,115, 110,120,115,120
- Depth 2:- 145, 140, 140,135,130, 135,140,135
- **Solution**: D:-15, 15,20,20,20,15,25,15 $D^{-} = \frac{\sum D}{N}$

D=145, D^{-} =18.13

$D^{2=}$ 225,225,400,400,400,225,625,225= 2725

$$=96.87 \sum D^2 - \frac{(\sum D)^2}{N}$$

$$\text{s.e.of } \bar{D} = \sqrt{\frac{\sum D^2 - \frac{(\sum D)^2}{N}}{N(N-1)}}$$

=1.31

t= D^{-}=18.13/1.31 t = 13.83

>t at 1% l.s at 7 d.f =(3.50)

We reject the null hypothesis. There is a significant difference in the soil core samples drawn from two depth levels.

F-test

Example 5:-Food supplements were grown under two experimental conditions. Two random samples of 11 and 9 food supplements show the sample (s.d) of their weights, 0.8 and 0.5 respectively.

Assuming that weight distributions are normal, test the hypothesis that the true variances are equal. (F tab 0.05 (10,8)d.f=3.35)

Solution:--H0:- Variances are homogeneous.

H1:- Variances differ significantly.

Here n1 = 11, n2 = 9,

sx = 0.8, sy = 0.5

Sx2 = (n1/n1-1) sx2

Sy2 = (n2/n2-1) sy2,

Sx2 = (11/10)* (0.8)2 = 0.704

Sy2 = (9/8)*(0.5)2 = 0.281

F = 0.704/0.281 = 2.50

F- calculated is less than F-tabulated

We accept our null hypothesis

Chi-square test

Example 6: From the point of view of timber, test whether two samples, sample A and sample B, differ significantly.(x^2 tab0.05=3.38, tab 0.01=6.64)

	High timber value	Low timber value
Variety A	350	205
Variety B	250	195

Solution: Ho: The difference in the value of the two samples is non-significant H1: Alternative

Sample	Hig h	Low	Total
A	350	205	555
B	250	195	445
Total	600	400	1000

$$x^2 = \sum_{i=1}^{n} \frac{[0i - Ei]^2}{Ei}$$

=4.84 (Significant at 5% l.s)

We reject our null hypothesis at 5% l.s

There is a significant difference between the samples at 5 % level of significance.

But what about 1% level of significance?.

Analysis of Variance (ANOVA)

The analysis of variance is the systematic algebraic procedure of decomposing the overall variation in the response variable in an experiment into different physical assignable components. The fundamental principle of analysis of variance (ANOVA) and its important technique was developed by Sir Ronald R. Fisher in 1918.

Analysis of variance (ANOVA) is a collection of statistical models and their associated procedures, in which, the observed variance in a particular variable is partitioned into components attributable to different sources of variation. In its simplest form, ANOVA

provides a statistical test of whether or not the means of several groups are all equal and therefore, generalises t-test to more than two groups. ANOVAs are helpful because they possess an advantage over a two sample t-test. Doing multiple two sample t–tests would result in an increased chance of committing type I error. For this reason ANOVAs are useful in comparing three or more means.

Suppose we have a factor with 't' levels (t treatments) with equal replications 'r and their response is Y_{ij}

Model is One way

$Y_{ij} = \mu + t_i + e_{ij}$

where ti is the t treatments

μ = overall mean

eij=error term

Model is Two way:

$Y_{ij} = \mu + A_i + B_j + e_{ij}$

Ai = one factor

Bj = second factor

The ANOVA enables us to know whether the variation due to a component factor is significantly more than the variance due to the experimental error or not. Hence, the mean sum of squares due to that component is nothing but variance due to the experimental component.

The interpretation of data based on analysis of variance (ANOVA) is valid only when the following assumptions are satisfied:

1. Additive Effects: Treatment effects and block (environmental) effects are additive.
2. Independence of errors: Experimental errors are independent.
3. Homogeneity of Variances: Errors have common variance.
4. Normal Distribution: Errors follow a normal distribution.

Also, the statistical tests t, F, z etc. are valid under the assumptions of independence and normality of errors. Any departure from these assumptions makes the interpretation based on these statistical techniques invalid. Therefore, it is necessary to detect the deviations and apply the appropriate remedial measures.

Additive Effects: The effects of two factors, say treatment replication, are said to be additive if the effect of one factor remains constant over all the levels of other factors. A hypothetical set of data from randomised complete block (RCB) design, with 2 treatments and 2 replications, with additives effects is given in Table 1.

Table 1

Treatment	Replication I - II		Replication effect I - II
A	190	125	65
B	170	105	65
Treatment Effect(A-B)	20	20	

Here, the treatment effect is equal to 20 for both replications and the replication effect is 65 for both treatments.

When the effect of one factor is not constant at all the levels of other factors, the effects are said to be non-additive. A common departure from the assumptions of additivity in biological experiments is one where the effects are multiplicative. Two factors are said to have multiplicative effects if they are additive only when expressed in terms of percentages. Table 2 illustrates a hypothetical set of data with multiplicative effects.

Table 2

Treatment	Replication		Replication Effects	
	I	II	I-II	100(I-II)/II
A	200 (2.3010)	125 (2.0969)	75 (0.2041)	60
B	160 (2.2041)	100 (2.000)	60 (0.2041)	60
Treat Effect(A-B)	40 (0.0969)	25 (0.0969)		
100(A-B)/B	25	25		

In this case, the treatment effect is not constant over replications and the replication effect is not constant over treatments. However, when both the treatment effect and replication effect are expressed in terms of percentage, an entirely different pattern emerges. For such a violation of assumptions, the Logarithm transformation is quite suitable. For illustration, the Logarithm transformation of data in Table 2 is given.

Normality of Errors

The assumption of the homogeneity of variance and normality are generally violated together.

To test the validity of normality of errors for the character under study, one can take the help of Normal Probability Plot, Anderson-Darling Test, D'Augstino's.

Test, Shapiro – WILK's Test, Ryan-Joiner test, Kolmogrov-Smirnov test, etc. In general, moderate departures from normality are of little concern in the fixed effects ANOVA as F- test is slightly affected. In the case of random effects, it is more severely impacted by non-normality. Significant deviations of errors from normality makes the inferences invalid. So before analysing the data, it is necessary to convert it to a scale so that it follows a normal distribution.

In the data form designed field experiments, we do not directly use original data for testing of normality or homogeneity of observations, because this is embedded with the treatment effects and some other effects like block, row, column etc. So, there is a need to eliminate these effects from the data before testing the assumptions of normality and homogeneity of variances.

Homogeneity of Error Variances

A crude method for detecting the heterogeneity of variances is based on scatter plots of means and variances, but Bartlett's test for homogeneity of variances is a valid one. If we have only two variances, then F statistics is used.

Remedial Measures: Data transformation is the most appropriate remedial measure in situations where the data is non-normal and variances are heterogeneous. The data is converted into a new scale resulting in a new data set that is expected to satisfy the homogeneity of variances. Depending upon the functional relationship between variances and means, suitable transformation is used.

The following are the three transformations, which are being used most commonly in research data (mostly biological).

Transformation of data

i) Logarithmic transformation

ii) Square root transformation

iii) Arc Sine transformation

iv) Reciprocal Square root

v) Reciprocal transformation

i) ***Logarithmic transformation:*** This transformation is suitable for the data where the variance is proportional to the square of the mean or the coefficient of variation (S.D./mean) is constant or where effects are multiplicative. These conditions are generally found in the data that are whole numbers and cover a wide range of values. This is usually the case when analysing growth measurements such as trunk girth , length of extension growth, weight of tree or number of insects per plot, number of egg mass per plant or per unit area etc.

For such situations, it is appropriate for the data sets to analyse log X instead of actual data X. (When the data set involves?)

For such situations, it is appropriate to analyze log X instead of actual data X.

When data set involves small values or zeros, log (X+1) or log (X+0.5) should be used instead of log X.

ii) ***Square-Root Transformation:*** This transformation is appropriate for the data sets where the variance is proportional to the mean. Here, the data consists of small whole numbers, for example, data obtained in counting rare events, such as the number of infested plants in a plot, the number of insects caught in traps, number of weeds per plot. This data set generally follows the Poisson distribution and square root transformation approximates Poisson to normal distribution.

For these situations, it is better to analyze $\sqrt{X}$ than that of the actual data X. If X is confirmed to small whole numbers, then, $\sqrt{X+1}$ or $\sqrt{X+0.5}$

This transformation is also appropriate for the percentage data, where the range is between 0 to 30% or between 70 to 100%.

iii) ***Arc Sine Transformation:*** This transformation data is appropriate for the data on proportions, i.e., data obtained from a count and the data expressed as decimal fractions and percentages. The distribution of percentages is binomial and this transformation makes the distribution nomial. Since the role of this transformation is not properly understood, there is a tendency to transform any percentage using arc sine transformation. But only that percentage data that are derived from count data, such as % barren tillers (which is derived from the ratio of the number of non-bearing tillers to the total number of tillers), should be transformed and not the percentage data such as %protein or %carbohydrates, % nitrogen, etc., which are not derived from count data. For these situations, it is better to analyse sin-1 ($\sqrt{X}$) than that of X, the actual data. If the value of X is 0%, it should be substituted by (1/4n) and the value of 100% by (100-1/4n), where n is the number of units upon which the percentage data is based.

It is interesting to note here that not all percentage data need to be transformed and if they do, arc sine transformation is not the only transformation possible. The following rules may be useful in choosing the proper transformation scale for percentage data derived from count data.

Rule 1: The percentage data lying within the range 30 to 70 % is homogeneous and no transformation is needed.

Rule 2: For percentage data lying within the range of either 0 to 30% or 70 to 100%, but not both, the square root transformation should be used.

Rule 3: For percentage data that does not follow the ranges specified in Rule 1 or Rule 2, the Arc Sine transformation should be used.

The other transformations used are **reciprocal square root** (when variance is proportional to the cube of mean), **reciprocal** (when variance is proportional to the fourth power of mean) and **tangent hyperbolic** transformation.

The transformations discussed above are particular cases of the general family of transformations known as Box-Cox transformation.

Box-Cox transformation:- If the relation between the variance of observations and the mean is known, this information can be utilised in selecting the form of the transformation. The transformation given by Box-Cox(1964) is a power transformation of the original data$_{ut}$.Let y_{ut} be the observation. Then

$$y^*_{ut} = y_{ut}^{\lambda}$$

The particular cases for different values of λ are given below

λ	Transformation
1	No transformation
½	Square root
0	Log
-1/2	Reciprocal square root
-1	Reciprocal

Even when the data remains non-normal or heterogeneous after transformation, then Non-parametric tests are used.

References

A Text Book of Agricultural Statistics by R. Ranga Swamy. Biostatistics by Wayne W. Daniel.
Design and Analysis of Experiments. Angela Dean and Daniel Voss
Handbook of Biological Statistics by John H. McDonald
Statistical Methods for Agricultural Workers by V.G.Panse and P.V. Sukhatme
Statistical Methods for Environmental and Agricultural Sciences by A. Reza Hoshmand
Statistical Procedures for Agricultural Research, 2nd Edition by Kwanchai A. Gomez & Arturo A. Gomez

Index

N

P

Q

R

S

T

W